Bioorganic Marine Chemistry

Volume 3

Edited by Paul J. Scheuer

With contributions by
A.R. Davis, M.P. Foster, C.M. Ireland,
J. Kobayashi, M. Kobayashi, O.J. McConnell,
T.C. McKee, T.F. Molinski, D.J. Newman,
Y. Ohizumi, D.M. Roll, K. Sakata, K. Snader,
M. Suffness, N. Suzuki, J.C. Swersey,
N.M. Targett, C.M. Young, T.M. Zabriskie

With 33 Figures

Springer-Verlag Berlin Heidelberg New York
London Paris Tokyo Hong Kong

Professor Paul J. Scheuer
University of Hawaii at Manoa, Department of Chemistry
2545 The Mall, Honolulu, Hawaii 96822, USA

ISBN 3-540-50870-8 Springer-Verlag Berlin Heidelberg New York
ISBN 0-387-50870-8 Springer-Verlag New York Berlin Heidelberg

Typesetting: Macmillan India Ltd., Bangalore-25;
Printing: Color-Druck, Dorfi GmbH, Berlin;
Bookbinding: Lüderitz & Bauer, Berlin

2151/3020-54210 – Printed on acid-free paper

Preface

The first three chapters of Vol. 3 of *Bio-organic Marine Chemistry* deal with the chemistry and function of peptides. Chapter 1 by Ireland and coworkers serves as an introduction to marine-derived peptides. It is arranged phyletically and encompasses the entire range from dipeptides to a compound with 95 amino acid residues. Peptides involved in primary metabolism and hence belonging to the realm of macromolecular biochemistry are excluded. However, it might be mentioned in passing that the dividing line between large and small molecule chemistry is continually becoming less distinct. Not only are more compounds of intermediate size, from 1,000 to 10,000 dalton, being discovered, but instruments and techniques, particularly in mass spectrometry and nuclear magnetic resonance have been developed for their structural elucidation by what is considered small molecule methodology.

Two groups of peptides are discussed in separate chapters. Biologists who have observed and described the mating behavior of diverse species of marine invertebrates have long surmised that a chemical mechanism might be operating in many cases of individual as well as mass fertilization. The chemical activators of sea urchin sperm prove to be a series of peptides, whose structures and activity are discussed by Suzuki.

The second distinct group of peptides reviewed in this volume are the cone shell toxins. Because of their potency and the ingenious delivery system which the animal uses, these toxins have intrigued biologists and chemists alike, not to mention writers of crime and passion. Kobayashi et al. discuss several cone shell toxins, whose structures have recently been elucidated.

Two chapters of Vol. 3 deal with ecological topics. Davis et al. lay the groundwork with a broad treatment of an important phenomenon in marine biology: epibiosis. Research in this area is particularly difficult since separation of the epibiont from the host creates a new system with its own relevancies. Hence it is gratifying that a number of well-documented examples have been studied at the molecular level.

The second ecology chapter by Sakata deals with a vital aspect of all life, feeding behavior. The author discusses a group of animals which include herbivores and carnivores, and draws attention to many sound biological observations with as yet few chemically based answers.

The final chapter by Suffness et al. of the U.S. National Cancer Institute is *sui generis*. The readers of this series are well aware that much of today's marine natural product research is performed not primarily for its ecological significance, but for its potential in biomedicine and agriculture. Suffness et al. once again make it abundantly clear that success in this latter endeavor rests not on the prowess of the chemist, but on the sophistication and accuracy of the bioassays and on the availability of the potential drug in sufficient quantity. The account of the twenty-year bryostatin-1 saga makes fascinating and instructive reading.

Once again it is my pleasure to acknowledge the cooperation of all contributors. As always, comments and suggestions from the readers are welcome.

November 1988 Paul J. Scheuer
June 1989

Table of Contents

Natural Product Peptides from Marine Organisms

Chris M. Ireland, Tadeusz F. Molinski, Deborah M. Roll, T. Mark Zabriskie, Tawnya C. McKee, J. Christopher Swersey, and Mark P. Foster[1]

Contents

Abstract

This chapter reviews the literature on peptides and modified peptides isolated from marine animals and plants that appear to be products of secondary metabolic pathways (i.e., do not appear to have a primary metabolic function). Although the focus is towards biologically active peptides, marine peptides with unique stuctures are also discussed, regardless of whether they have reported biological activity. Polypeptides, e.g., enzymes, structural proteins and receptor molecules will be specifically excluded from this discussion, as they are more appropriate for a biochemistry review article. For the purpose of this review, peptides will be defined as compounds derived biosynthetically from the condensation of at least two amino acids to produce a peptide (amide) bond, which may or may not have undergone further modification.

Over the past thirty years thousands of marine natural products have been reported, proving marine organisms to be a rich and varied source of new structural classes of secondary metabolites, many without precedent in terrestrial natural products. Interest in this field has grown steadily: in just an eight year period between 1977 and 1985, approximately 1700 new compounds were reported from marine organisms. The compounds discussed in this review are organized by phylum beginning with those from algae, invertebrates and concluding with those from vertebrates.

1 Department of Medicinal Chemistry, University of Utah, Salt Lake City, Utah 84112

Bioorganic Marine Chemistry, Volume 3

1 Introduction

This chapter reviews the literature through 1987 on peptides and modified peptides isolated from marine animals and plants that appear to be secondary metabolites (i.e., do not appear to have a primary metabolic function). The literature describing peptides isolated from the cone shell *Conus* spp. and the echinoderms will not be covered because they are the subject of separate chapters in this volume. However, peptides from freshwater algae, particularly the cyanobacteria, will be included. Although the focus is towards biologically active peptides, marine peptides with unique structures are also discussed, regardless of whether they have reported biological activity. Polypeptides (e.g., enzymes, structural proteins and receptor molecules) will be specifically excluded from this discussion, as they are more appropriate for a biochemistry review article. However, for completeness a number of peptides from sea anemones, heteronemertine worms and fish will be discussed. For the purpose of this review, peptides will be defined as compounds derived biosynthetically from the condensation of at least two amino acids to produce a peptide (amide) bond, which may or may not have undergone further modification. Several conventions have been adopted in this discussion. Absolute configurations

Table 1. Amino acid shorthand

Amino acid	Three-letter abbreviation	One-letter symbol
Alanine	Ala	A
Arginine	Arg	R
Asparagine	Asn	N
Aspartic acid	Asp	D
Carboxyglutamic acid	Gla	Z
Cysteine	Cys	C
Glutamine	Gln	Q
Glutamic acid	Glu	E
Glycine	Gly	G
Histidine	His	H
Hydroxyproline	Hyp	X
Isoleucine	Ile	I
Leucine	Leu	L
Lysine	Lys	K
Methionine	Met	M
Phenylalanine	Phe	F
Proline	Pro	P
Serine	Ser	S
Threonine	Thr	T
Tryptophan	Trp	W
Tyrosine	Tyr	Y
Valine	Val	V

follow the convention of Cahn, Ingold and Prelog. Peptide sequences of greater than 20 residues will be depicted by one or three letter notations (see Table 1). In cases where a compound has been reported by different names, the name first cited in the literature will take precedent.

The birth of the branch of organic chemistry now called Marine Natural Products is generally credited to Bergmann who in 1950 isolated the modified arabino nucleosides spongouridine and spongothymidine from a *Tedania* sponge [1]. These compounds, which served as models for synthesis of the clinically important drugs Ara-A and Ara-C, generated a great deal of interest in marine organisms as a potential source of pharmaceutical agents. Over the past 30 years thousands of marine natural products have been reported, proving marine organisms are rich and varied source of new structural classes of secondary metabolites, many without precedent in terrestrial natural products. Interest in this field has grown steadily: in just an eight year period between 1977 and 1985 approximately 1700 new compounds were reported from marine organisms [2].

Despite excitement generated by sponge nucleosides, the early literature was dominated by reports of nonnitrogenous metabolites. These included halogenated terpenes and fatty acid derivatives from red algae and sea hares, cembranes and eicosanoids from coelenterates, sesquiterpenes from sponges and nudibranchs, and diterpenes from brown algae. [3, 4, 5]. The paucity of nitrogenous metabolites reported from marine organisms in the early literature was often attributed to the fact that nitrogen is a limiting nutrient in the ocean. In retrospect, it is now clear that the choice of study organisms was at least equally important. For example, 40% or more of the secondary metabolites isolated from sponges, marine microbes, bryozoans, tunicates and blue-green algae (Cyanophyta) in the last 10 years contain nitrogen [2]. In contrast, less than 10% of those reported from red algae, brown algae, coelenterates, molluscs and echinoderms are nitrogenous.

A diverse array of nitrogenous metabolites including nucleosides, alkaloids [6], isonitriles, guanidines [7], amino acids [8, 9, 10] and peptides have been isolated from marine organisms. Several of these topics have been the subject of specific review articles, and nitrogenous metabolites in general have been covered in comprehensive reviews of the field by Faulkner [11, 12, 13] and Krebs [14].

The compounds discussed in this review are organized by phylum of the producing organism. The order of presentation is arbitrary, beginning with algae, followed by invertebrates and concluding with vertebrates.

2 Algae

While only a few simple peptides have been isolated from red (Rhodophyta), brown (Phaeophyta), and green (Chlorophyta) algae, blue-green algae (Cyanophyta) have proven to be a rich source of peptides with diverse structure and biological activity. The peptides from macroalgae have been reviewed [8, 9, 10, 15, 16].

Carnosine (**1**), or *β*-alanyl-(*S*)-histidine, a substance ubiquitous in animal tissue, was detected in the rhodophyte *Acantophora dendroides* and six other species of red

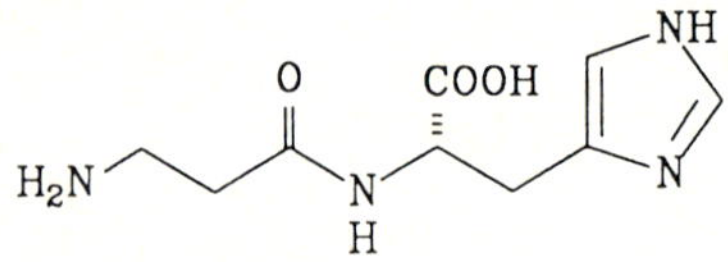

1

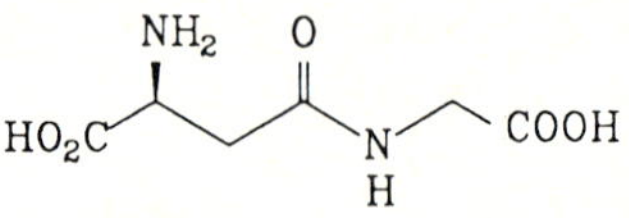

2

algae by chromatographic techniques [17, 18]. *Ceramium rubrum* yielded β-aspartyl-glycine (**2**) in a yield of 0.22% based on wet weight [19]. Methanol extracts of *Grateloupia turutura* gave, after repeated ion-exchange chromatography, (*S*)-citrullinyl-(*S*)-arginine (**3**) [20]. Several analogous oligopeptides containing citrulline, arginine and ornithine have recently been reported from *G. livida* [21]. Extraction of an aqueous extract of *G. livida* has also yielded the dipeptide (*S*)-gigartinyl-(*S*)-gigartinine (**4**) [22].

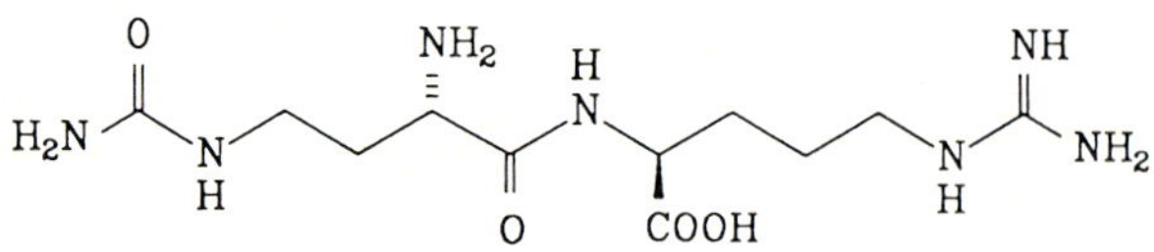

3

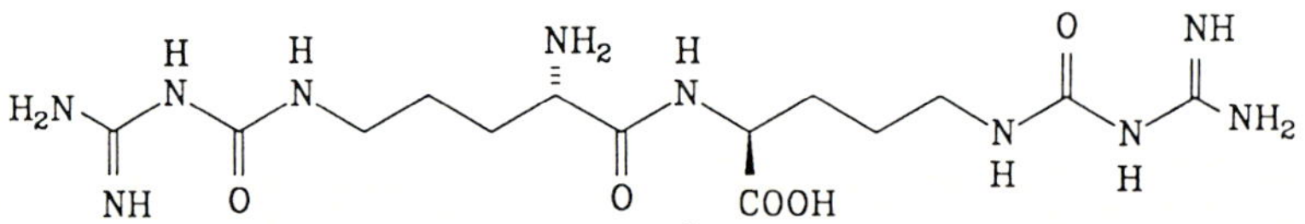

4

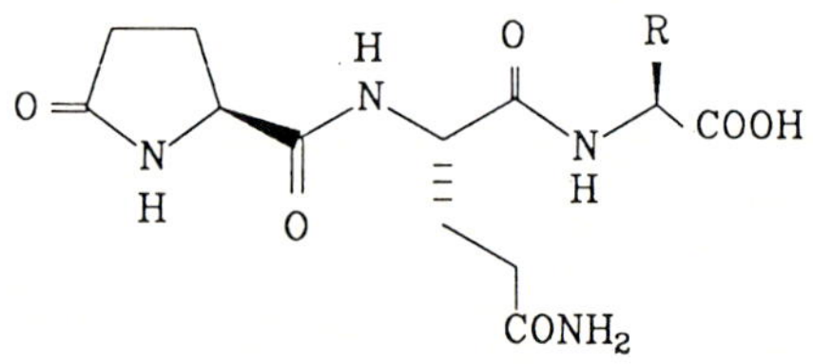

5 R = CH_3

6 R = $(CH_2)_2CONH_2$

Eisenine (**5**) ((*S*)-pyroglutamyl-(*S*)-glutaminyl-(*S*)-alanine) was isolated from the phaeophytes *Eisenia bicyclis* [23] and *Ecklonia cava* [24]; the structure proposed on the basis of chemical reactions [23, 25, 26, 27, 28] and confirmed by synthesis [29, 30, 31]. A related peptide, fastigiatine (**6**) from *Pelvetia fastigiata* [32], was formulated as (*S*)-pyroglutamyl-(*S*)-glutamyl-(*S*)-glutamine; however, the reported melting point and optical rotation of the isolated compound do not coincide with that of a synthetic peptide with the proposed structure [30, 31]. It is postulated that the pyroglutamyl residues of both eisenine and fastigiatine may be artifacts of isolation arising from cyclization of a glutaminyl residue in the respective native peptides. The brown alga *Analipus japonicus* gave the incompletely characterized pentapeptide analipine, comprised of glutamic acid and aspartic acid or their amides [33].

Extraction of the fresh water green alga *Cladophora* sp. gave (*S*)-arginyl-(*S*)-glutamine (**7**) [34]. This same peptide was also detected in the marine species *Enteromorpha linza* and isolated from *Ulva lactuca* [35].

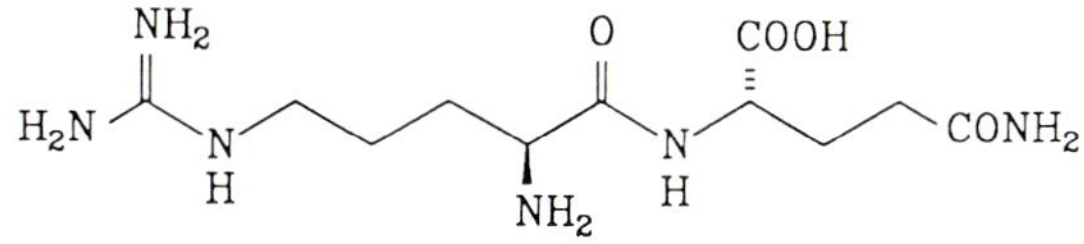

7

Despite considerable work reported on simple peptides from macroalgae, little of their biological activity is known. Indeed, the metabolic function of peptides in algae deserves a satisfactory explanation as these compounds, together with free amino acids, are often found as major nitrogenous constituents of algal tissue.

Blooms of the cosmopolitan freshwater blue-green alga *Microcystis aeruginosa* in eutrophic potable waters of ponds and reservoirs have been implicated in the death of livestock and also human intoxication [36, 37, 38, 39, 40, 41, 42]. The causative agents, a family of cyclic heptapeptides, are powerful hepatotoxins which cause dose dependent chronic liver damage leading to death. The structures of five of these compounds were determined by a collaborative effort involving groups in England, South Africa, and Australia [37, 43, 44]. Toxic material was isolated from cultured strains of algae obtained from samples of toxic blooms in South Africa. Further purification of this toxic extract afforded four related cycloheptapeptides, cyanoginosins LA (**8**) (formerly known as toxin BE-4) LR (**9**), YR (**10**) and YA (**11**). A fifth peptide, cyanoginosin YM (**12**), was obtained from a toxic bloom from New South Wales, Australia [44, 45]. The structures were obtained by extensive NMR analysis together with fast atom bombardment mass spectrometric (FABMS) studies of the intact peptide and of the products of partial hydrolysis. Each peptide contains *erythro*-*β*-methyl-(*R*)-isoaspartate and the new amino acid 3-amino-9-methoxy-2,6,8-trimethyl-10-phenyldeca-4,6-dienoic acid of undefined stereochemistry. The toxins proved quite potent; one study reported an LD_{50} for cyanoginosin YM of 0.056 mg/kg for mice injected intraperitoneally [46]. Cyanoviridin RR (**13**) is one of three cyclic peptides obtained from an axenic culture of *M. viridis* (NIEA-102) collected in Japan [47, 48]. The sequence was determined largely by NMR including heteronuclear multiple bond correlation (HMBC) and homonuclear Hartmann-Hahn (HOHAHA) spectroscopy. The stereochemistry was not determined.

Many recent reports have appeared describing the isolation and characterization of hepatotoxic cyclic peptides from strains of *Microcystis aeruginosa.* This literature is somewhat confusing, as in most cases the complete structure and sequence have not been determined and it is not clear whether these report re-isolations of known cyanoginosins or new structures.

Certain varieties of *Lyngbya majuscula* (Oscillatoriaceae), a toxic marine cyanophyte found in warm waters, have been implicated as the causative agents of

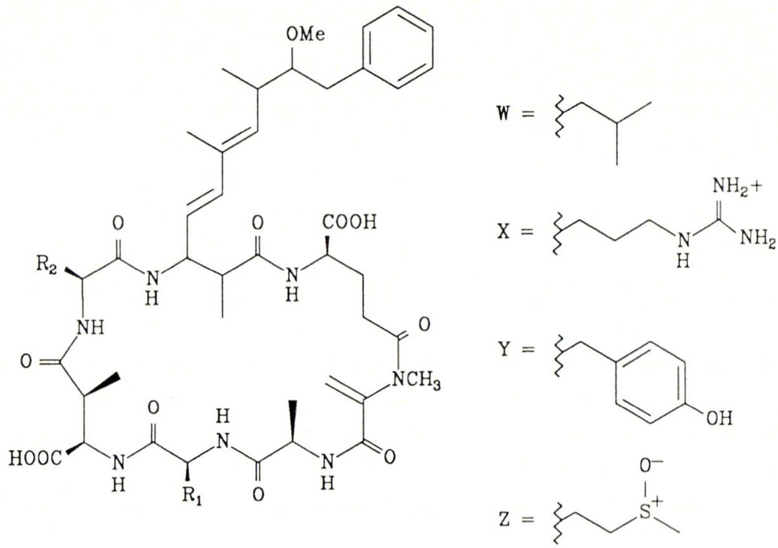

8 R_1 = W, R_2 = Me
9 R_1 = W, R_2 = X
10 R_1 = Y, R_2 = X
11 R_1 = Y, R_2 = Me
12 R_1 = Y, R_2 = Z
13 R_1 = X, R_2 = X

"swimmers' itch," a form of contact dermatitis [49]. Several workers reported isolation of a lipophilic toxin from collections of the algae [50, 51, 52], and a purified fraction was shown to elicit dermonecrosis at a dose of 0.08 μg when injected into mice [49]. Moore and his colleagues at the University of Hawaii succeeded in isolating a potent irritant from *L. majuscula* and defined the gross structure as the prenylated cyclic dipeptide lyngbyatoxin-A (**14**) by spectroscopic methods [52]. The configuration of the dipeptide moiety of **14** was established by chiroptical comparison with the closely related teleocidin B (**15**), a *Streptomyces*

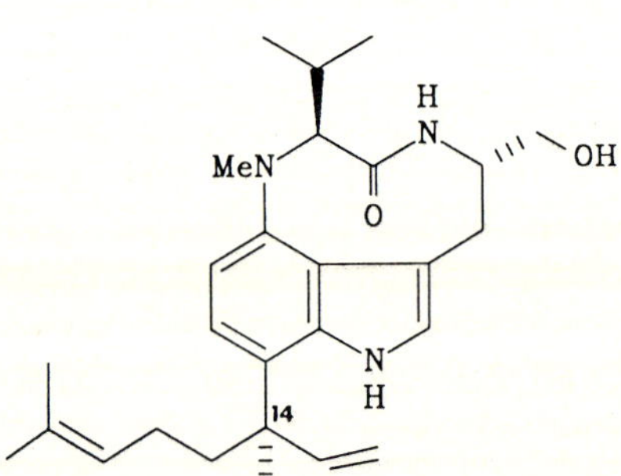

14

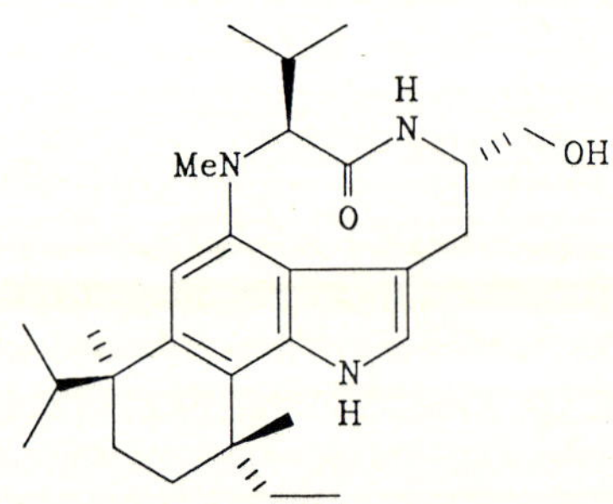

15

metabolite [53]. The configuration of the C-14 quaternary carbon was subsequently shown to be (*R*) by degradation of **14** to dimethyl 2(*S*)-2-ethyl-2-methylglutarate [54]. Teleocidin A-1, recently isolated from the terrestrial bacterium *Streptomyces mediocidicus* [55, 56], is identical with lyngbyatoxin-A, and congener teleocidin A-2, is the C-14 epimer of **14** [54]. Lyngbyatoxin-A and the teleocidins represent a new structural class of potent tumor promoters that exhibit tumor promoting activity in mice similar to that induced by phorbol esters [57]. Lyngbyatoxin-A was found to kill baitfish (*Poecilia vittata*) at a concentration of 0.15 μg/mL in seawater, suggesting *Lyngbya majuscula* may produce lyngbyatoxin-A as an ichthytotoxic chemical defense [52]. Lyngbyatoxin-A and teleocidin A-2 have recently been synthesized [58].

The majusculamides A (**16**), B (**17**), C (**18**) were isolated from two different varieties of *L. majuscula*. Majusculamides A and B, nontoxic peptides obtained from a shallow water form of the alga [59], are the C-2 epimers of the lipodipeptide *N*-(2-methyl-3-oxodecanoyl)-(*R*)-*N*,*O*-dimethyl-tryosyl-(*S*)-*N*-methylvalinamide. These compounds were not present in a second deep water variety of the same alga

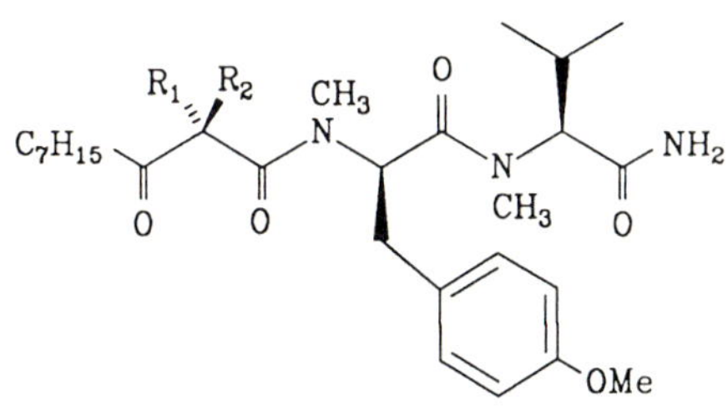

16 R_1 = H, R_2 = Me

17 R_1 = Me, R_2 = H

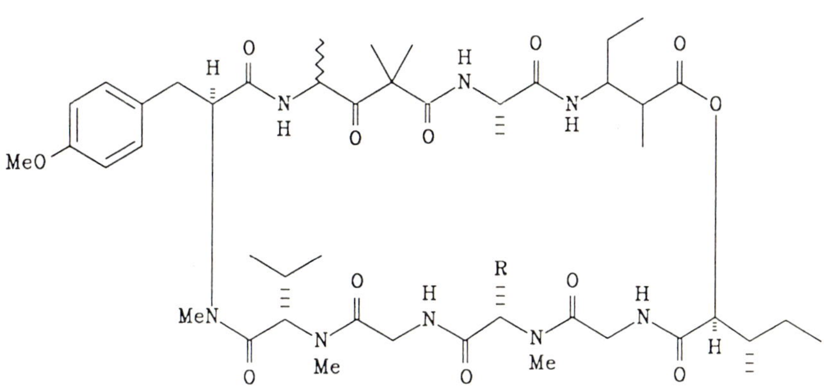

18 R = 2-Mepropyl

156 R = *i*-butyl

which instead contained the antifungal depsipeptide majusculamide C [60]. The composition of **18** was established by analysis of hydrolysis products and the sequence deduced by analysis of mass spectral fragmentation and ^{1}H NMR nuclear Overhauser effect (nOe) measurements. Compound **18** is comprised of several unusual amino acids including 3-amino-2-methylpentanoic acid of unknown stereochemistry and (*S*)-*N*,*O*-dimethyltyrosine (opposite absolute configuration to that of majusculamides A and B). Neither of the majusculamides are responsible for the toxicity of the alga, which is attributed to lyngbyatoxin-A (**14**), aplysiatoxin and debromoaplysiatoxin.

Large-scale cultivation of blue-green algae for isolation of bioactive metabolites offers several advantages over classical harvesting of wild varieties. Biologically active compounds, obtainable from collected blue-green algae in otherwise vanishingly small amounts, may be produced on a practical scale in the laboratory. This was certainly a factor in solving the structures of the cyanoginosins. The ability to control the growth of an organism in a "closed" environment simplifies biosynthetic studies employing isotope labelling and also opens the possibility of producing modified natural products by incorporation of synthetic precursors in the growth medium.

The puwainaphycins A (**19**), B (**20**), C (**21**), and D (**22**) were isolated from a cultured strain of the freshwater blue-green alga *Anabaena* sp. [61]. The puwainaphycins contain a number of uncommon amino acids, the most striking of which

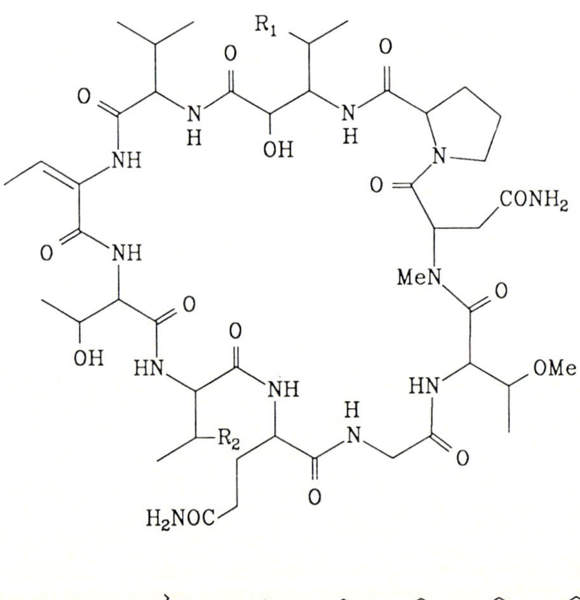

19 R_1 = X, R_2 = OH
20 R_1 = X, R_2 = Me
21 R_1 = Y, R_2 = OH
22 R_1 = Y, R_2 = Me

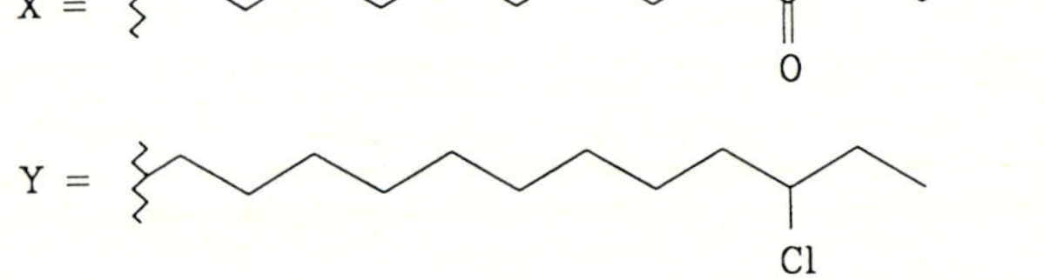

are the fatty acid derived amino acid, 3-amino-2-hydroxy-4-methyl-14-oxostearic acid in **19** and **20** and a rare chlorine-containing amino acid, 3-amino-14-chloro-2-hydroxy-4-methylpalmitic acid in **21** and **22**. The structures were solved by a combination of classical hydrolytic amino acid analysis, mass spectrometry and ^{1}H- and ^{13}C-NMR techniques including the use of two-dimensional COSY and relayed coherence transfer (RCT). Puwainaphycin A was shown to exhibit potent cardiotonic activity.

Application of 2D ^{1}H-NMR techniques (COSY, NOESY, multiple quantum filters, etc.) [62] to the study of peptide composition, sequence and solution conformation is relatively new and presents some advantages over classical techniques. Modified peptides comprised of nonprotein amino acids or containing unusual sidechains or "end caps" are often not amenable to classical hydrolytic or enzymatic degradation alone. Two-dimensional ^{1}H NMR provides a powerful nondestructive method for obtaining nonempirical structural information for the elucidation of peptide structures.

3 Sponges

Over the past 20 years, the Porifera have proven to be a source of an extraordinary number of metabolites, the majority of which are nonpeptidal. Most sponge peptides are severely modified, so that only a few "true" peptides exist which possess recognizable amino acid linkages.

The first antimicrobial peptide to be characterized from a sponge was discodermin A (**23**), isolated from *Discodermia kiiensis* collected in the Izu Archipelago [63,64,65]. Discodermin A inhibited the growth of *Bacillus subtilis*, *Staphylococcus aureus*, *Escherichia coli*, and *Pseudomonas aeruginosa* at 1 μg/disk. The tetradecapeptide structure of **23** was determined by Edman degradation of the desformyl compound (Ala-Phe-Pro-**t**-Leu-**t**-Leu-Trp) and analysis of the BNPS-skatole reaction product (H-Arg-Cys(O_3H)-Thr-MeGln-Leu-Asn-Thr-Sar). Electron impact mass spectrometry (EIMS) fragmentation patterns established the structures of discodermins B-D (**24-26**), which differ only in the fourth and fifth residues from the *N*-terminus [65]. In addition to possessing antimicrobial activity, discodermins A and C inhibited the development of starfish embryos at 5 μg/mL. The **t**-Leu residues in discodermin A had been found previously only in Actinomycete metabolites. Since *Discodermia* is known to contain algal and

HCO-(R)-Ala-(S)-Phe-(S)-Phe-(S)-Pro-X-(R)-Trp-(S)-Arg-(R)-Cys(O_3H)-(S)-Thr-(S)-MeGln-(R)Leu-(S)-Asn-(S)-Thr-Sar

23 X=(R)-t-Leu-(S)-t-Leu
24 X=(R)-Val-(S)-t-Leu
25 X=(R)-t-Leu-(S)-Val
26 X=(R)-Val-(S)-Val

CH_3OCH_2C-(S)-Val-(R)-Me-Leu-(S)-Thr-X-(R)-Leu-(S)-Me-Ile-β-Ala-(R)-aIle-(S)-Me-Val-(S)-Me-Ala-(S)-β-Ala-(R)-Leu-(R)-Me-aIle

27 X=β-ala

bacterial symbionts, it has been suggested that the discodermins are symbiont products [63].

Five novel cyclic depsipeptides were isolated from an Okinawan *Theonella* sp. [66]. Acid hydrolysis of the major component, theonellapeptolide Id (**27**), yielded 13 amino acids, five of which contained an *N*-methyl group, and revealed that the *N*-terminal valine unit was protected as a methoxyacetyl group. The sequence of **27** was determined by acidic and enzymatic hydrolyses, analysis of FABMS data, and characterization of the hydrolysates as dansyl derivatives. The ester linkage in threonine was identified by the ^{1}H-NMR signal of the β-hydrogen in several hydrolysate fragments. The absolute configuration of the amino acids, five of which were (*R*), was secured by circular dichroic (CD) analysis of the acid hydrolysates. Theonellapeptolide Id inhibits the development of fertilized sea urchin eggs at 50 μg/mL. The structures of the four remaining minor peptides have not yet been elucidated. However, they inhibit urchin development at much lower concentrations-2, 2, 2, and 10 μg/mL, respectively. This same tridecadepsipeptide was independently reported from a *Theonella* sp. collected at Okinawa Island, and named theonellamine B. In this latter report the structure was established by analysis of 2D NMR chemical shift correlations of hydrolysis products, making extensive use of long-range coupling information. Theonellapeptolide Id is the first peptide reported to be a specific inhibitor of Na, K-ATPase (IC_{50}, pig brain Na, K-ATPase, 7×10^{-6} M) [67]. In 1987 Kitagawa reported the structure of theonellapeptolide Ie (**28**), which differs from **27** in *N*-methylation of one of the β-alanine units [68].

28 X = Me-β-ala

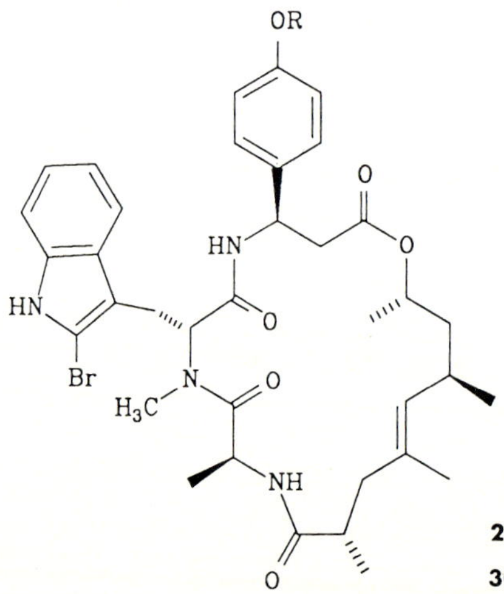

29 R = H

30 R = Ac

Jaspamide (**29**), a unique depsipeptide having a wide range of biological activities, was independently isolated from *Jaspis* sp. collected in Fiji and Palau [69]. Jaspamide contains three amino acids and a twelve-carbon propionate-derived unit. It exhibited potent insecticidal activity (LC_{50}.4 ppm) against the tobacco budworm *Heliothis virescens* and inhibited the growth of *Candida albicans* at 1 μg/disk. The alanine, β-tyrosine, and 2-bromoabrine (2-bromo-*N*-methyl tryptophan) residues were assigned on the basis of 2D ^{1}H-^{13}C NMR data. X-ray analysis of a tyrosyl acetate derivative **30** unambiguously established both the structure and relative stereochemistry of jaspamide. The absolute configuration was assigned on the basis of chiral HPLC of hydrolysis products. Subsequently, the same compound was reported under the name jasplakinolide [70]. In this paper, jaspamide was reported to exhibit an in vitro ED_{50} of less than 1 μg/mL against the nematode *Nippostrongylus braziliensis*, in vitro cytotoxicity of 0.32 μg/mL against larynx epithelial carcinoma and 0.01 μg/mL against human embryonic lung cells. Jaspamide displayed in vivo activity comparable to miconazole nitrate when administered topically as a 2% solution to a *Candida* vaginal infection in mice [70]. Jaspamide has also been isolated from *Jaspis johnstoni* collected in New Guinea [71].

The geodiamolides A and B (**31**–**32**) were isolated from *Geodia* sp. collected in the West Indies. The geodiamolides are 18-membered ring cyclic depsipeptides, which differ only in the substitution of a halogen in the *ortho* position of *N*-methyl tyrosine. The structures and stereochemistry of both geodiamolides were solved by X-ray analysis. Interestingly, they possess the same 12-carbon propionate-derived unit as jaspamide. Also like jaspamide, the aromatic amino acid bears an *R* absolute configuration. The geodiamolides were reported to be mildly antifungal with an MIC of 31.3 μg/mL against *Candida albicans* [72].

Two additional amino acid derivatives have been isolated from an unidentified Fijian sponge of the Jaspidae family. Bengamides A (**33**) and B (**34**) are novel

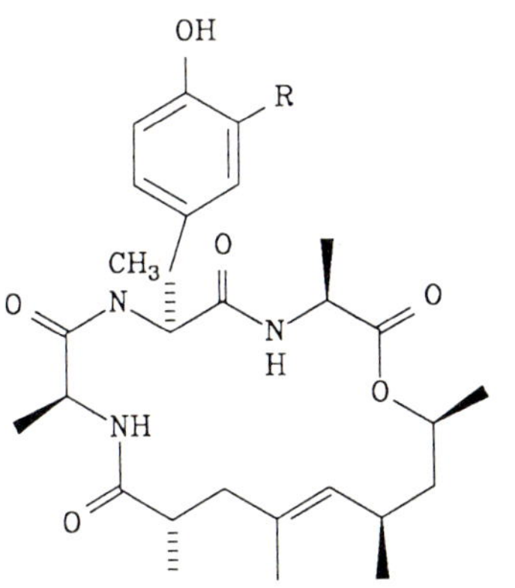

31 R = I

32 R = Br

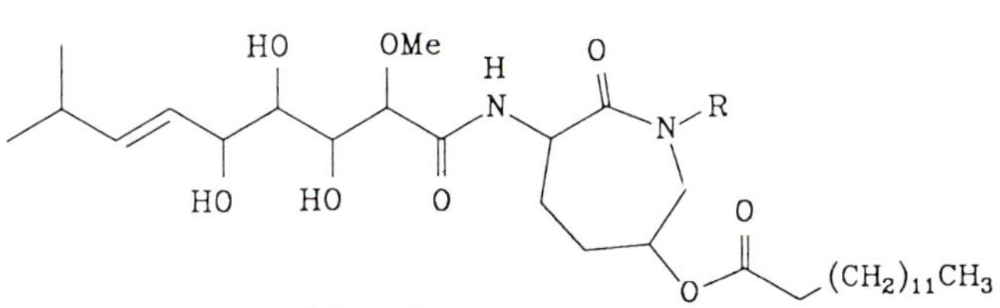

33 R = H

34 R = CH_3

seven-membered ring heterocycles cyclized by a δ-hydroxylysine. The structures of **33** and **34** were largely established using 2D ^1H-^1H and ^1H-^{13}C COSY and long-range NMR coupling experiments. Bengamides A and B were both active against *N. braziliensis* at 50 μg/mL and *Streptococcus pyrogenes* at 3.9 and 1.9 μg/mL, respectively [73].

Clionamide (**35**) is the major metabolite of a British Columbian burrowing sponge, *Cliona celata*, that has the ability to bore into the calcium carbonate shells of barnacles and rock scallops [74, 75]. Due to the instability of **35** and difficulty in obtaining pure material, the crude extract was acetylated and characterization completed on the tetraacetyl derivative **36**. Ozonolysis of **36** yielded 3,4,5-triacetoxybenzaldehyde and treatment with methanolic H_2SO_4 gave *N*-acetyl-6-bromotryptophan methyl ester (**37**), confirming the structure of **36**. Treatment of tetraacetyl clionamide with HCl/CH_3CN and subsequent hydrolysis provided **38**. Comparison of the optical rotation of **38** with literature values established (*S*) absolute stereochemistry at C-2. Compound **36** and hydrogenation product **39** have also been synthesized [75, 76]. Clionamide was mildly antibiotic against *S. aureus*. The ability of catechol groups, such as those present in clionamide, to chelate calcium suggests a possible role for this compound in the ability of *Cliona* to excavate coral, cited as a major source of coral reef erosion [75].

Four other peptide alkaloids related to clionamide have been isolated from *C. celata*. Celenamides A (**40**) and B (**41**), and the minor metabolites C (**42**) and D (**43**) were all characterized as acetyl derivatives **44**–**47** [77, 78]. The celenamides are the

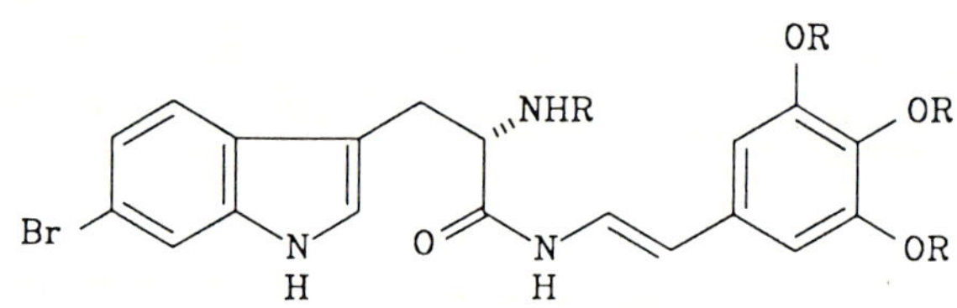

35 R = H

36 R = Ac

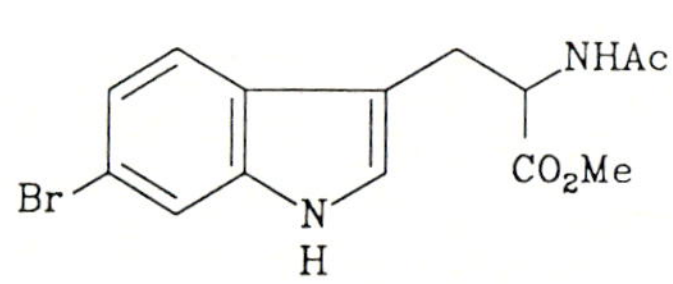

37

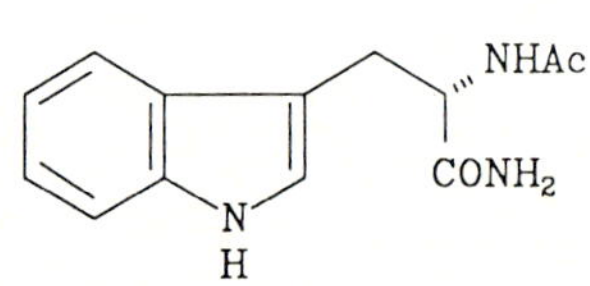

38

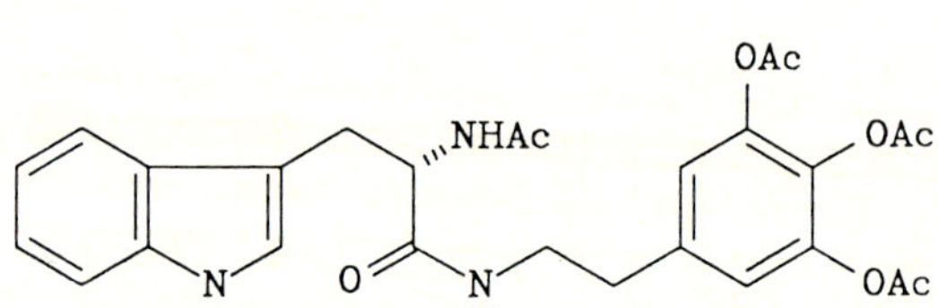

39

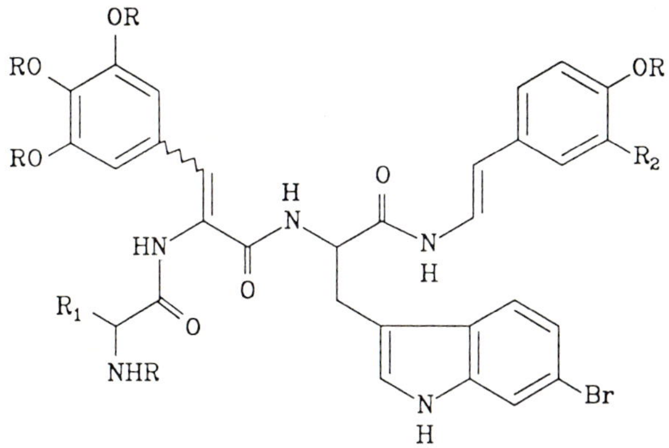

40 R = H, R_1 = iBu, R_2 = OH

41 R = H, R_1 = iPr, R_2 = OH

42 R = H, R_1 = iBu, R_2 = H

44 R = Ac, R_1 = iBu, R_2 = OAc

45 R = Ac, R_1 = iPr, R_2 = OAc

46 R = Ac, R_1 = iBu, R_2 = H

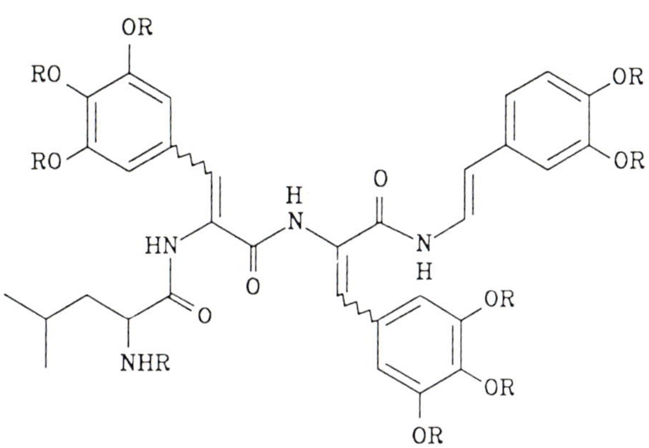

43 R = H

47 R = Ac

first nonterrestrial members of a peptide alkaloid family exemplified by the plant products integerrin (**48**) and lasoidine-A (**49**). They possess *N*-methylated amino groups, C-terminal *trans* geometry, and a novel amino acid, characterstics without precedent in this family of metabolites. The recent synthesis of hexaacetyl celenamide A (**44**), by the condensation of phosphonoglycine esters with a 6-bromotryptophan derivative, represents the first total synthesis of a linear peptide alkaloid [79].

Certain sponges have demonstrated strict conformity to chemotaxonomic patterns, especially those in the order Verongida. Every species examined from

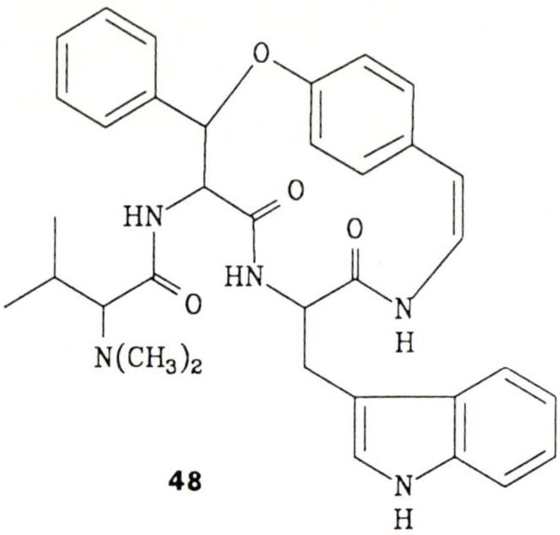

48

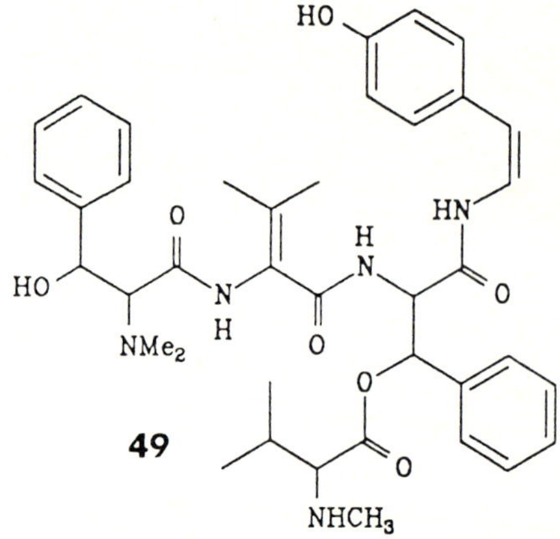

49

this order produces secondary metabolites derived from a brominated tyrosine. The bastadins, which contain four bromotyrosine units, were reported from the Australian sponge *Ianthella basta* [80, 81]. Bastadins -1 (**50**), -2 (**51**) and -3 (**52**) are acyclic components derived from a single phenolic oxidative coupling reaction, while bastadins -4 (**53**), -5 (**54**), -6 (**55**), and -7 (**56**) contain a 28-membered macrocyclic ring formed from two oxidative couplings. All seven compounds displayed potent in vitro and some in vivo antimicrobial activity against Gram positive organisms [81]. The molecular formulas of the metabolites were established from high resolution mass measurements on their methyl derivatives. The presence of oxime functionalities and the amino acid sequence of bastadin-1 were determined largely by the analysis of EI mass spectral data and fully supported by ^{1}H NMR. Hydrolysis of the pentamethyl derivative **57** followed by methylation gave 2-(3-bromo-4-methoxyphenyl)-ethylamine and the dimethyl ester **58**, which confirmed the oxidative coupling between the central bromotyrosine rings. Bastadins -2 and -3 were identified in an analogous manner.

Bastadin-5 (**54**) was isolated only as the tetramethyl ether **59** and its structure unambiguously determined by X-ray analysis. Compound **54** is a macrocycle resulting from an additional phenol oxidative coupling of bastadin-2 (**51**). The spectral data of bastadin-6 (**55**) were similar to that of **54** and indicated that **55**

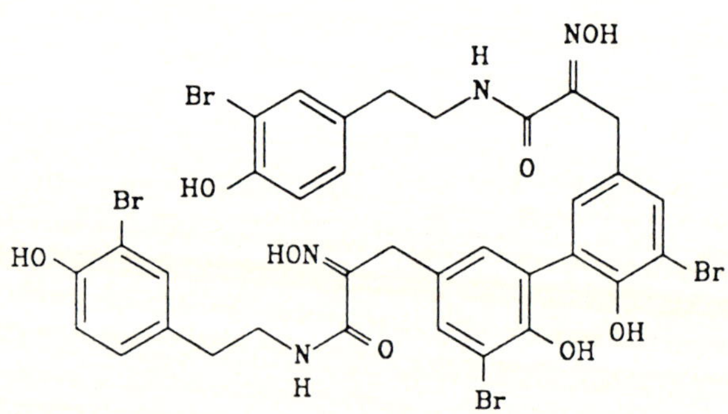

50

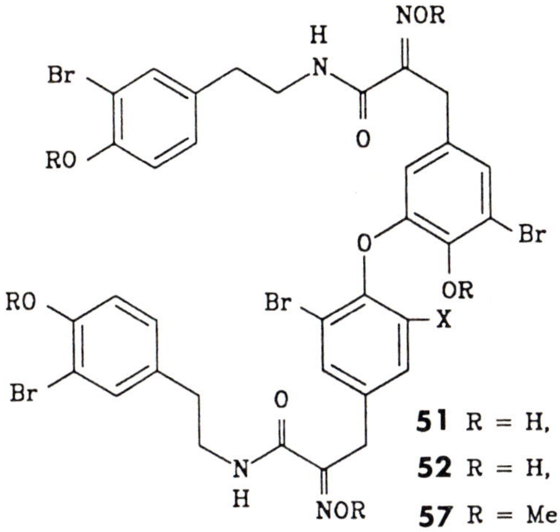

51 R = H, X = H
52 R = H, X = Br
57 R = Me, X = H

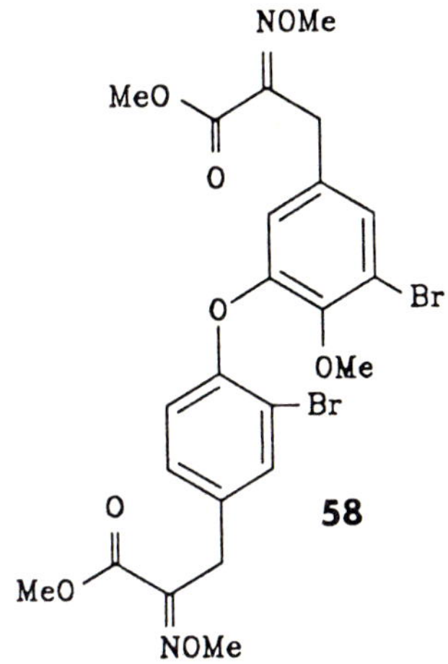

58

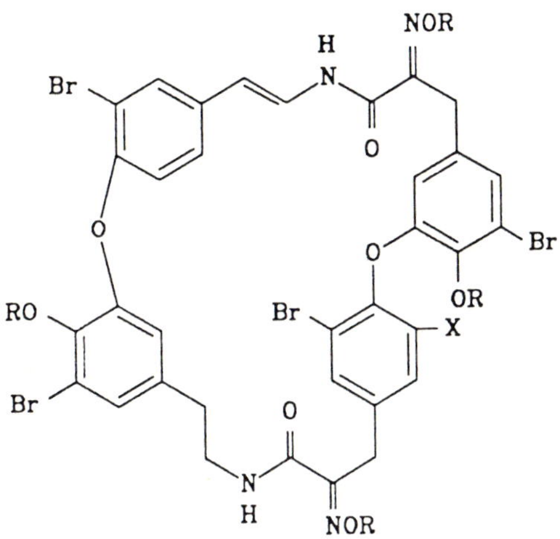

53 R = H, X = Br
56 R = H, X = H

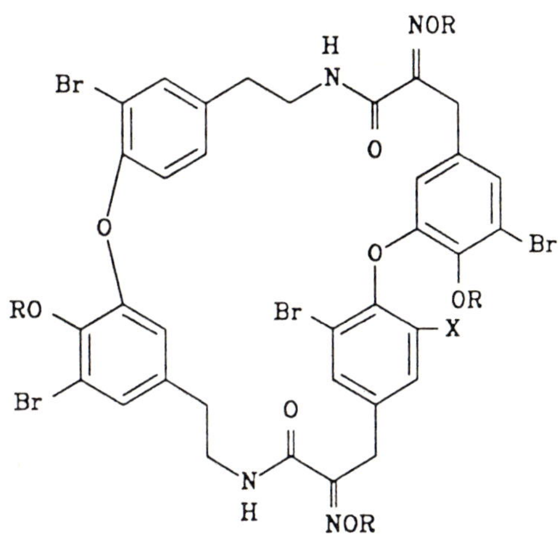

54 R = H, X = H
55 R = H, X = Br
59 R = Me, X = H
60 R = Me, X = Br

contained an extra bromine atom and two symmetrically substituted aromatic rings. Bastadin-4 (**53**) is a didehydro derivative of bastadin-5. The last member in the series, bastadin-7 (**56**), was shown by analysis of ^1H NMR and mass spectral data to be a debromo derivative of bastadin-4. Bastadins -1, -2, and -3 have been synthesized via a biomimetic type coupling of a 3,5-dibromotyrosine derivative to a 3-bromotyramine moiety [82]. The macrocyclic bastadin-6 tetramethyl ether (**60**) has also been synthesized by phenol oxidative coupling of precursor bastadin-2 catalyzed by thallium nitrate [83].

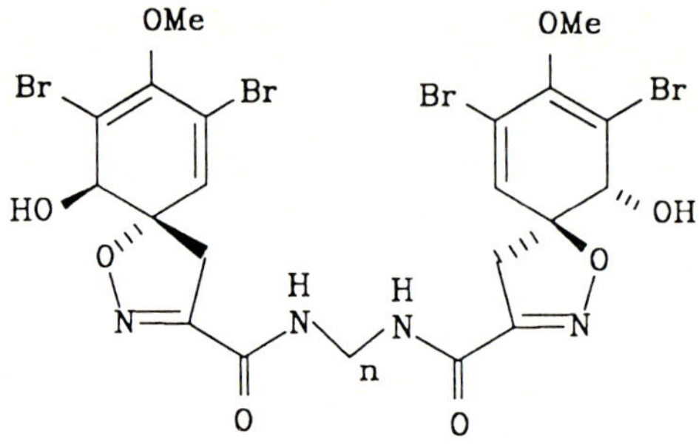

61 n = 4

62 n = 5

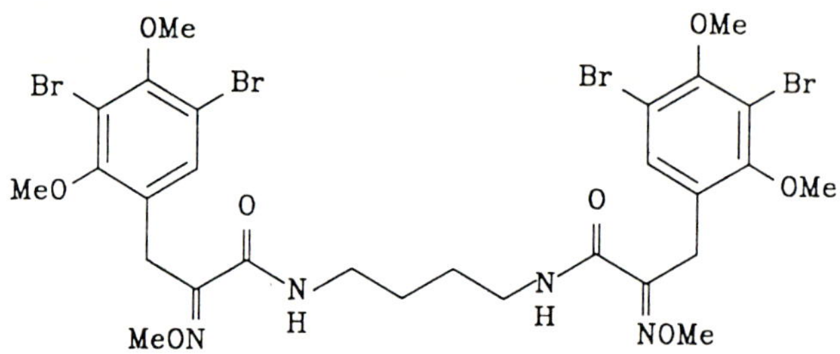

63

Aerothionin (**61**) and homoaerothionin (**62**) were the first complex spirocyclohexadienylisoxazole derivatives isolated from a verongid sponges [84, 85]. The major metabolite **61** has now been isolated from several species of *Aplysina* (*Verongia*) including *A. aerophoba* (later reidentified as *A. cavernicola*) [86] and *A. fistularis* (*V. thiona*), and may constitute up to 10% of the crude sponge extract [87]. The symmetrical nature of aerothionin was evident from the simplicity of the ^{1}H NMR spectrum; however, since the compound was optically active, the two halves of the molecule were not mirror images. Treatment of **61** with base cleaved the spiral ring forming an oxime that was characterized as the hexamethyl ether **63**. Aerothionin and homologue **62** appear to be derived from the condensation of 3,5-dibromotyrosine with ornithine and lysine, respectively. The spiral ring could arise by either the nucleophilic attack of an oxime group on an arene epoxide or conversion of an arene epoxide into a phenol with subsequent intramolecular phenol-oxime coupling [85]. The structure of aerothionin has now been confirmed by X-ray crystallographic analysis and its absolute configuration determined by comparison of CD spectra to those of (–)-aeroplysinin-1 (**64**) [87]. Both **61** and **62** have been synthesized in racemic form by condensation of the spiro compound **65** with 1,4-diaminobutane and 1,5-diaminopentane, respectively [88]. Electron probe microanalysis established that aerothionin and homoaerothionin are stored in the spherulous cells of *Aplysina fistularis* [89]. This is the first reported use of this

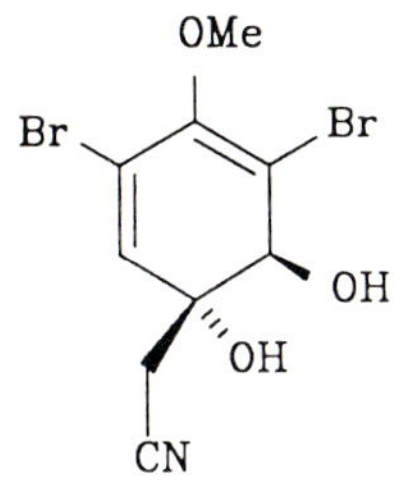

64

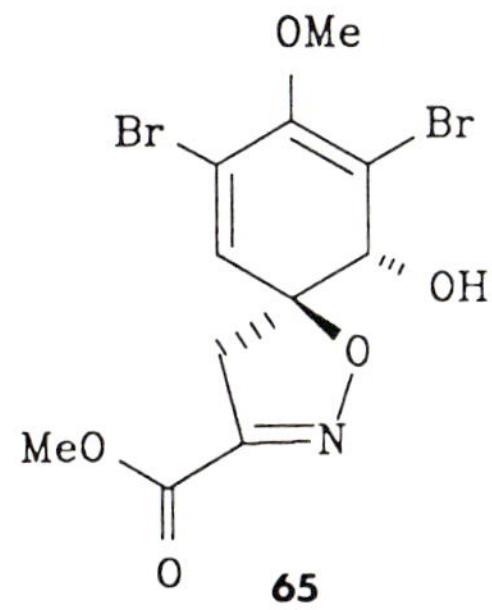

65

technique to localize a secondary metabolite at the cellular or subcellular level in a marine invertebrate. Electron micrographs indicate that the spherulous cells migrate to the excurrent canals of the sponge where they lyse. Release of these metabolites into seawater is proposed to be a defence mechanism against unwanted bacteria.

Recently, two additional bromotyrosine derivatives **66**–**67** with a central cystine unit were reported from an unidentified sponge collected in Guam. Compound **67** is reported to isomerize to **66**, leading to speculation that only the *E,Z* isomer is a natural product. Although the extract of the sponge was cytotoxic towards the P388 lymphocytic leukemia cell line at 0.7 μg/mL, neither **66** or **67** proved active [90]. Two additional groups have reported the isolation of **66** under the structure

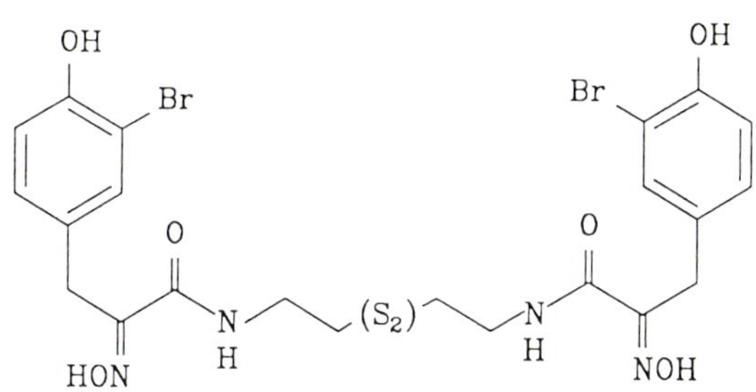

66 (*E,E*) oxime

67 (*E,Z*) oxime

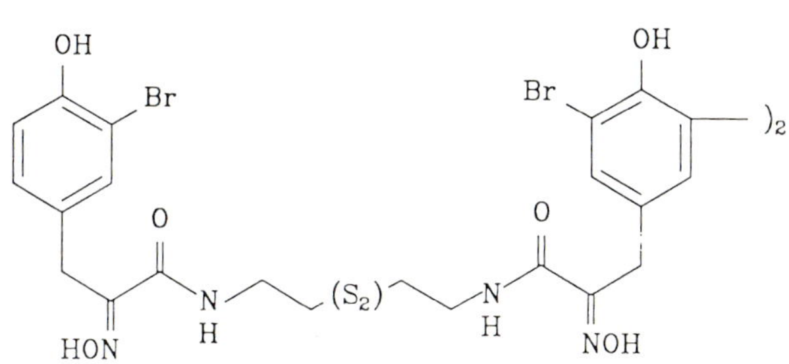

68

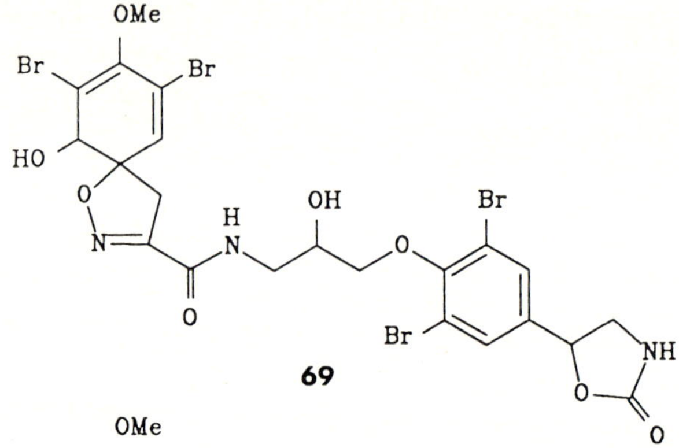

69

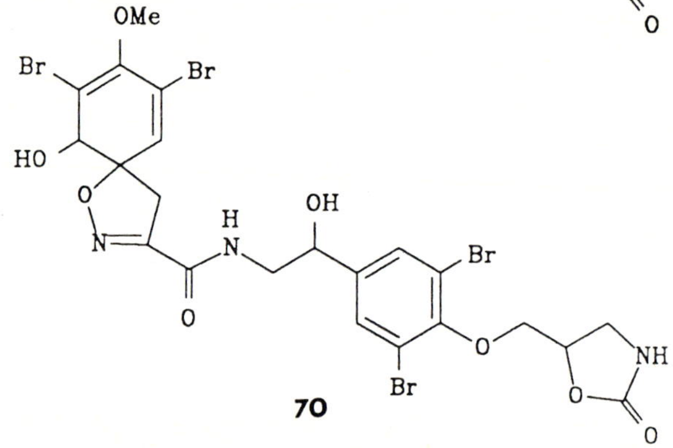

70

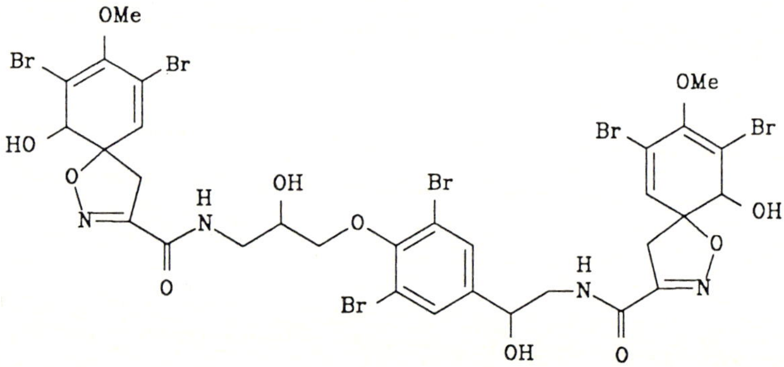

71

names, psammaplin A [91] and psammaplyn A [92]. In the former citation, **66** is reported to have an IC_{50} of 0.3 μg/mL against P388 and in the latter, antimicrobial activity. The latter paper also reports the isolation of a biphenyl dimer of **66**, bisaprasin **68**.

Four related metabolites were reported from extracts of sponges belonging to the genus *Aplysina*. Fistularins -1 (**69**), -2 (**70**), and -3 (**71**) were isolated from *A. fistularis* forma *fulva* collected in the Virgin Islands [93]. Isofistularin-3, from an Italian *A. aerophoba*, reportedly differs from **71** in stereochemistry at one or more chiral centers [86]. The stereochemistry of all of these compounds is as yet undefined. The structures of **69**–**71** were readily established by spectral analysis and detailed comparison of ^{1}H NMR data for the known *Aplysina* metabolites aerothionin (**61**) [85], and the bis-2-oxazolidone derivative **72** [94]. Fistularins -1 and -2 are composed of two dibromotyrosine and a modified β-alanine or threonine unit. Fistularin-3 (**71**) contains two spirocyclohexadienylisoxazole moieties but lacks the bis-2-oxazolidone functionality of **69** and **70**. Treatment of fistularin-3 and isofistularin-3 acetates with base followed by reacetylation gave the same hexaacetyl derivative **72** in accord with the chemistry observed for aerothionin. Both fistularin-3 and isofistularin-3 exhibited an ED_{50} of 4 μg/mL against KB cells [93]. Compound **71** also displayed in vitro cytotoxic activity against PS and LE cell lines at 4.3 and 1.3 μg/mL, respectively. Compound **72** was less active, inhibiting PS cells at 14 μg/mL. Fistularin-1 and a diacetate derivative also showed very mild in vitro cytotoxicity with ED_{50}'s of 21–35 μg/mL against KB, PS, and LE cell cultures [93].

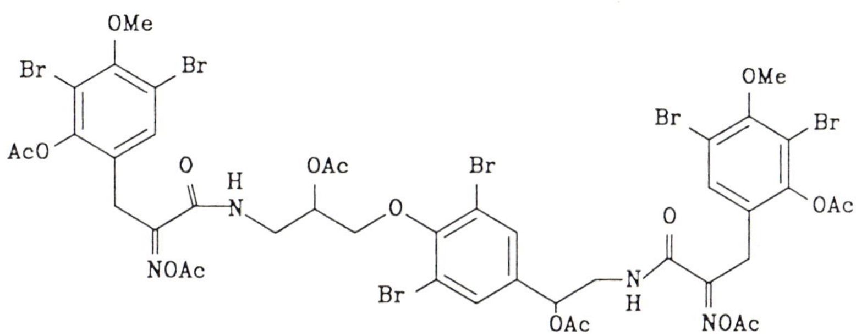

72

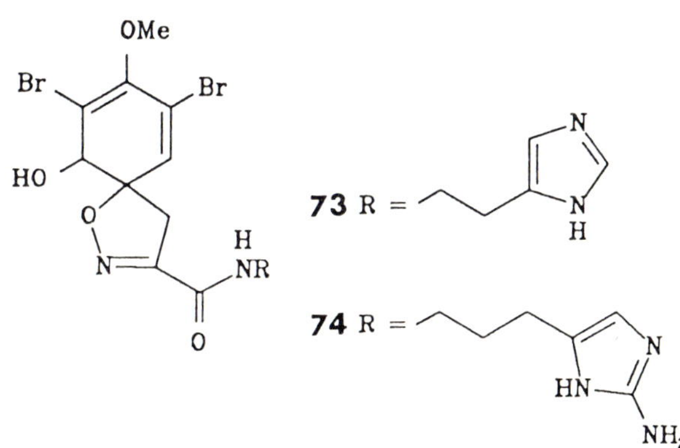

Compounds containing a dibromotyrosine unit linked to a histamine residue have been isolated from sponges collected in Italy, Okinawa, and the Caribbean. *Aplysina aerophoba* produces two such polar metabolites, aerophobins -1 (**73**) and -2 (**74**) [86]. ^{1}H NMR studies of **73** and **74** confirmed the presence of the same spirocyclohexadienylisoxazole ring system present in aerothionin (**61**), Acid hydrolysis of aerophobin-1 and -2 afforded histamine and 2-aminohomohistamine, respectively, confirming the differences in the two structures.

The Okinawan sponge *Psammaplysilla purea* produces purealin (**75**) which exhibits a notable inhibitory effect on Na, K-ATPase and myosin, Ca-ATPase [95]. The assignment of structure **75** was based principally on the analysis of 2D one-bond and long-range ^{1}H-^{1}H and ^{1}H-^{13}C NMR coupling data and chemical shift considerations. Purealin is composed of four amino acid residues – two dibromotyrosines, a modified aspartic acid, and a terminal 2-amino-homohistamine, arranged to form the same spirocyclohexadienylisoxazole system as in aerothionin (**61**), plus an aromatized oxime as found in bastadin-2 (**51**). The absolute configuration was established by comparison of the CD spectra of

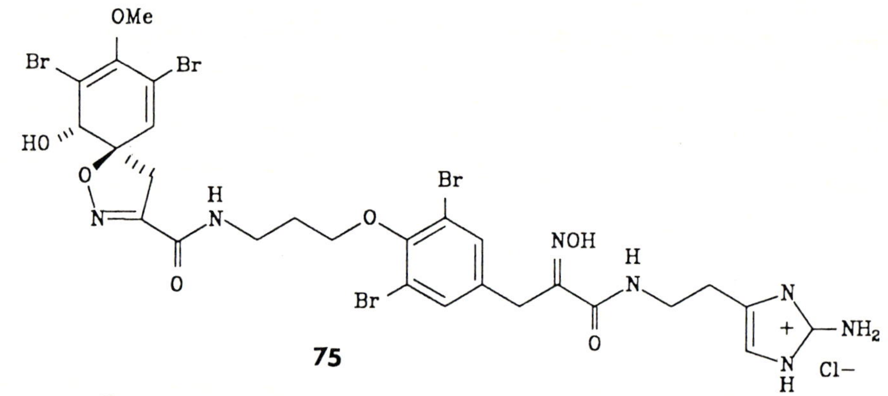

75

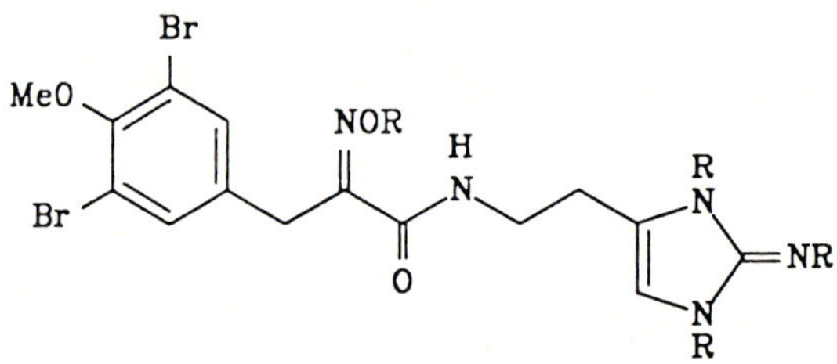

76 R = H

77 R = Ac

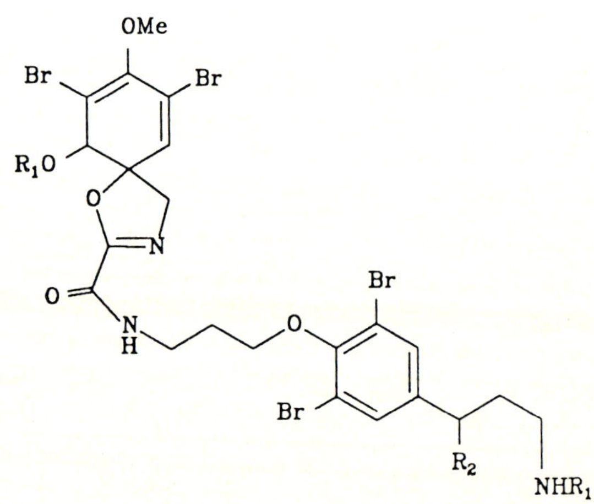

78 $R_1 = R_2 = H$

79 $R_1 = H$, $R_2 = OH$

purealin and aerothionin [87, 95]. Although **75** inhibits some ATPases, it was also the first natural product to activate K, EDTA-ATPase [95].

The most recent histamine containing peptide to be characterized from a verongid sponge is ianthelline (**76**), isolated from *Ianthella ardis*, collected in the Bahamas [96]. Spectral analysis of **76** and tetraacetate **77** revealed symmetrically substituted aromatic and aminoimidazole rings similar to those found in purealin [95] and aerophobin-2 (**74**) [86]. Ianthelline displayed only modest activity against *S. aureus* (15-mm zone of inhibition at 50 μg/disk) [96].

The last two bromotyrosine peptides to be discussed here are the psammaplysins -A and -B isolated from the sponge *Psammaplysilla purpurea*, collected in the Gulf of Aqaba and Palau [97, 98, 99]. The psammaplysins were originally assigned structures **78** and **79** on the basis of elemental composition, ^{1}H-NMR and degradation studies [97]. These compounds were anomalous because they represented the only examples of spiroxazolines (proposed to arise from a rearranged tyrosine) instead of the spiroisoxazoline normally found in verongid sponge metabolites. Subsequent reinvestigation of the psammaplysins by field desorption mass spectrometry (FDMS) established that psammaplysin-A diacetate (**80**) has a formula which includes one oxygen more than originally suggested by elemental analysis. Structure **80** was explicitly assigned to the diacetate on the basis of a ^{13}C-^{13}C 2D INADEQUATE NMR experiment; the structure and stereochemistry of **80** were also defined by X-ray analysis [99]. Therefore, psammaplysins -A and -B have structures **81** and **82**, respectively. The psammaplysins are derived from bromotyrosine through benzene oxide-oxepin intermediates to give a ring enlarged spiro ketal that retains the isoxazoline moiety found in all related verongid compounds. Both psammaplysins -A and -B exhibited in vitro activity against Gram-positive bacteria and *E. coli* [97, 98, 99].

The sponge *Dysidea herbacea* is known for its prolific production of pentabromophenols, sesquiterpenoids, and chloromethylated metabolites derived from

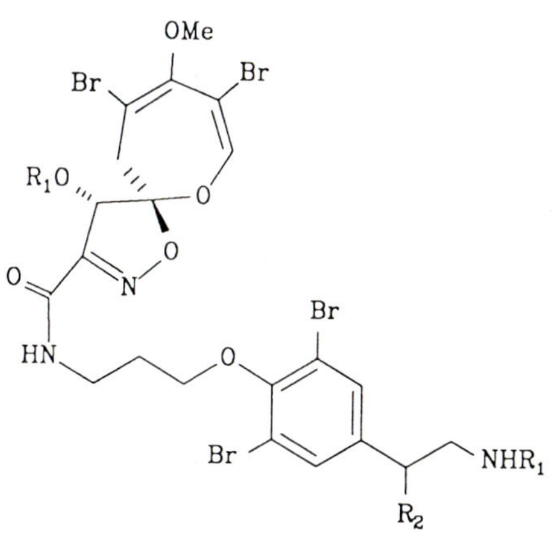

80 R_1 = Ac, R_2 = H

81 R_1 = R_2 = H

82 R_1 = H, R_2 = OH

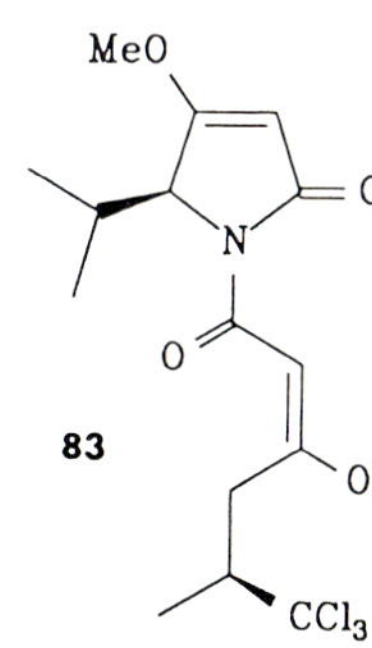

83

amino acids [100]. The tetramic acid derivative dysidin (**83**) was isolated from *D. herbacea* collected along the Great Barrier Reef near Townsville, Australia. Although not a true peptide, it contains the first naturally occurring trichloromethyl group [101]. The structure was determined by a combination of spectroscopic and degradation studies. X-ray analysis established both the relative and absolute stereochemistry. Synthesis of racemic dysidin confirmed the assignment as shown in **83** [102]. A sample of *D. herbacea* collected near Gladstone, Australia yielded the *N*-methyltrichloroleucine diketopiperazine derivative **84** [103]. The hexachloro derivative dysidenin (**85**) was the major metabolite from a specimen of *D. herbacea* collected near Cooktown, Australia [104]. Presence of the *N*-acylated trichloroleucine residue in **85** was established by ^1H and ^{13}C NMR spectroscopy while the 2-(2-thiazolyl)-alanine was supported by UV and mass spectral fragmentation data. Zinc/acetic acid reduction of dysidenin to the dechloro compound **86** provided further proof of structure **85**. *D. herbacea* from New Guinea produces isodysidenin (**87**), the C-5 isomer of **85** [105]. Treatment of **87** with diborane/THF followed by exhaustive N-methylation gave the crystalline quaternary amine **88**. X-ray analysis of **88** established the absolute configuration of isodysidenin as 2(*R*), 5(*S*), 7(*R*), 13(*R*). Spectral comparison of the acid hydrolysates of dechloroisodysidenin (**89**) corroborated the epimeric relationship of dysidenin and isodysidenin at C-5 [106]. Four additional members of this series were identified from *D. herbacea* collected along the Barrier Reef near Bowen, Australia [107]. The structures and stereochemistry of 13-demethyldysidenin (**90**), 13-demethylisodysidenin (**91**), 11-monodechloro-13-demethylisodysidenin (**92**), and 9-monodechloro-13-demethylisodysidenin (**93**) were assigned by interpretation of mass spectral fragmentation patterns and comparison of ^1H and ^{13}C NMR data and optical

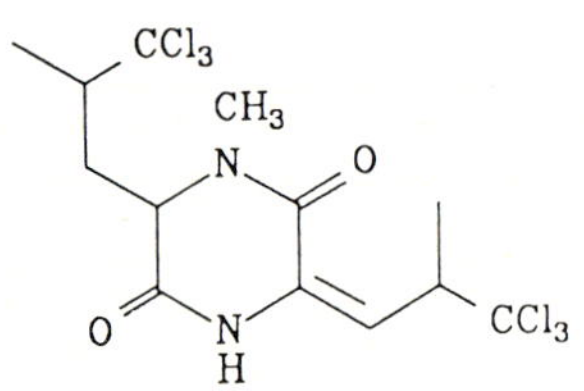

84

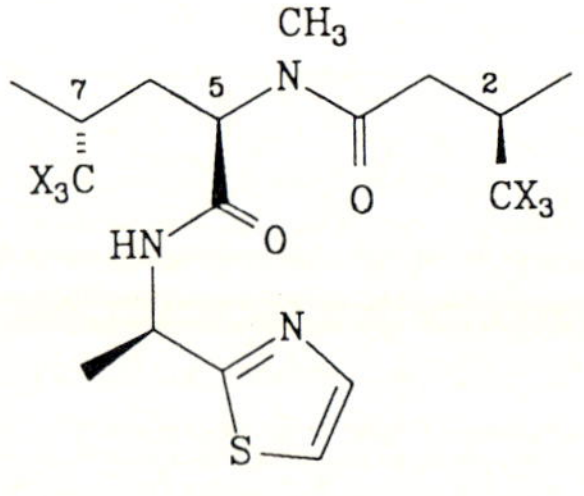

85 X = Cl

86 X = H

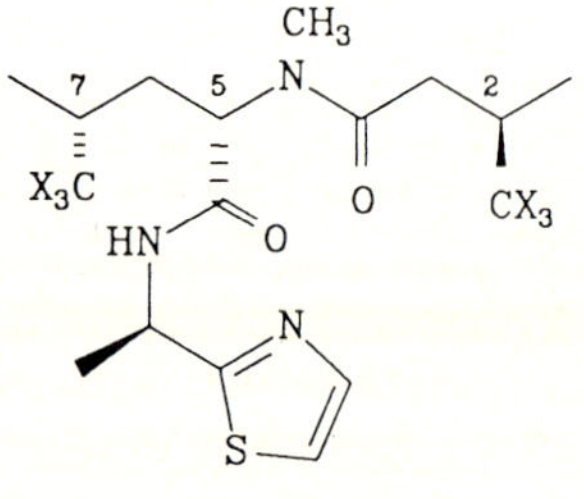

87 X = Cl

89 X = H

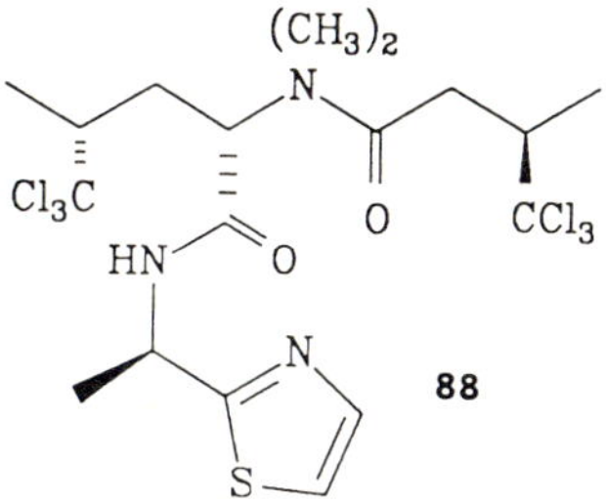

88

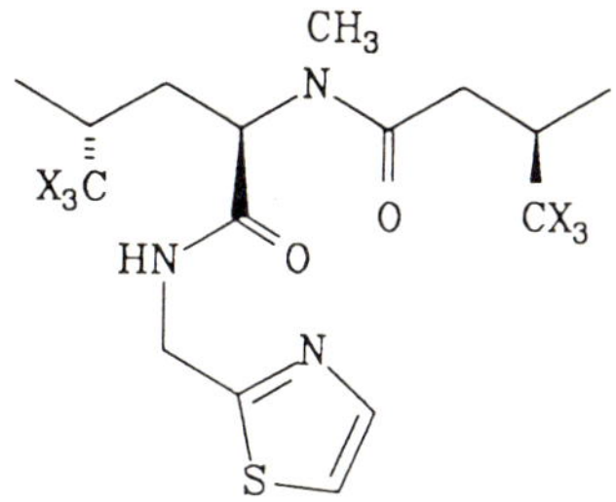

90 X = Cl

94 X = H

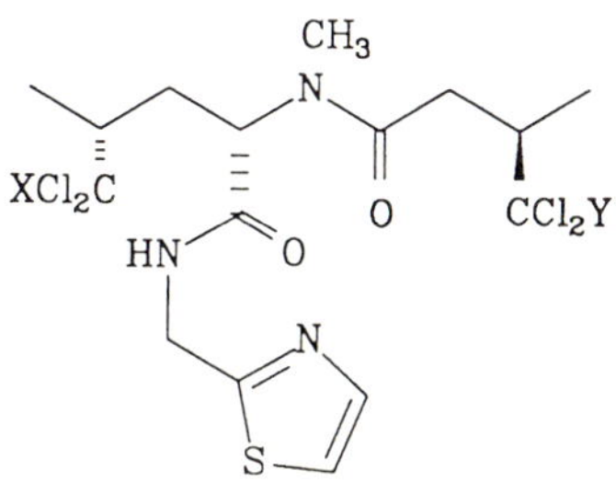

91 X = Y = Cl

92 X = Cl, Y = H

93 X = H, Y = Cl

95 X = Y = H

rotations with those reported for **85** and **87**. Both **90** and **91** were dechlorinated to yield compounds **94** and **95**, respectively, which differed only in their specific rotations, again due to the single asymmetric center at C-5.

The absolute configurations of dysidenin and isodysidenin were recently revised based on the results of a new method for establishing the chirality of 2-(1′-α-aminoalkyl) thiazoles. Alanine obtained by this method was assigned an (*S*) configuration, indicating the thiazole should also be reassigned an (*S*) configuration not (*R*) as determined by X-ray analysis [108]. In addition, the total synthesis of (+)-2(*R*), 5(*R*), 7(*R*)-13-demethyldysidenin (**96**) and (−)-2(*R*), 5(*S*), 7(*R*)-13-dimethylisodysidenin (**97**) confirmed the naturally occurring compounds possessed the opposite configuration from that previously reported [109]. Therefore, the absolute configuration of dysidenin is 2(*S*), 5(*S*), 7(*S*), 13(*S*); structure **98** and the related compounds are revised accordingly. Isodysidenin, the only member of this series to demonstrate ichthyotoxic activity, exhibits an LD_{50} of 5 mg/mL against *Lebistes*

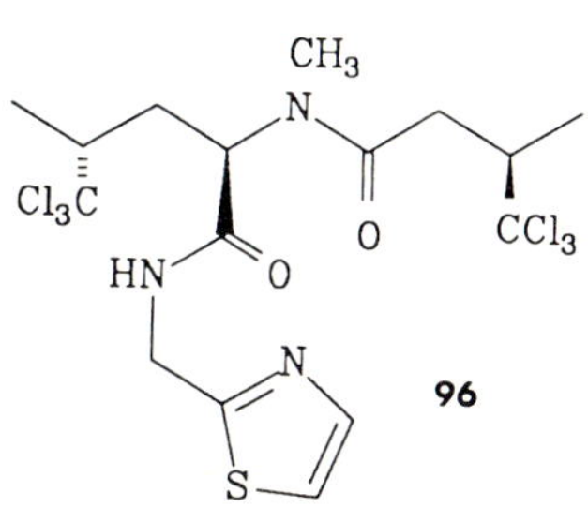

96

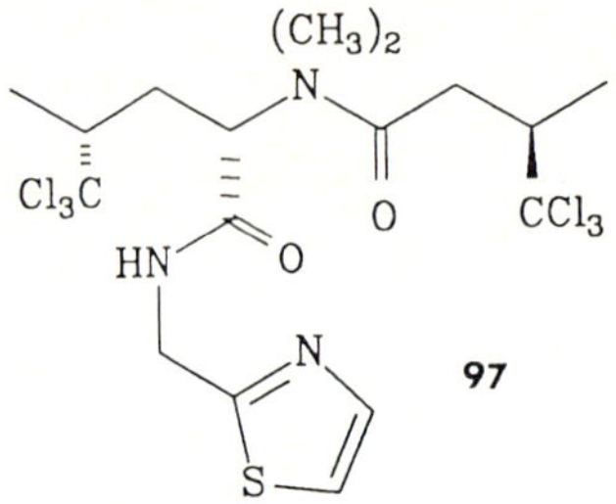

97

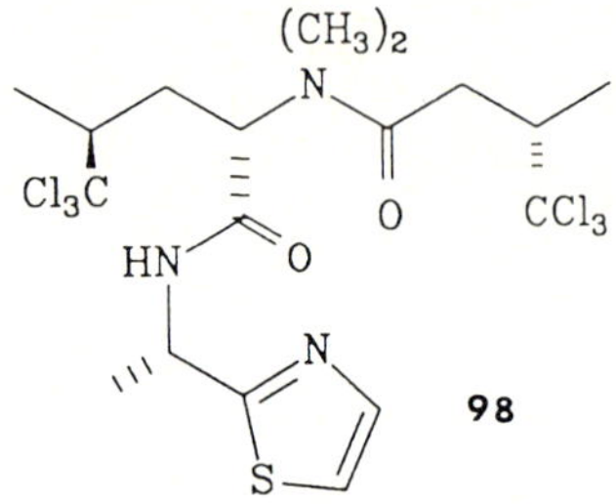

98

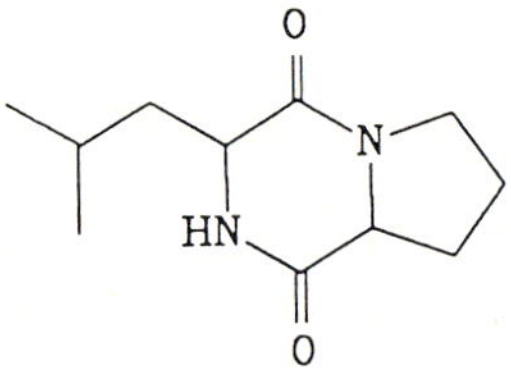

99

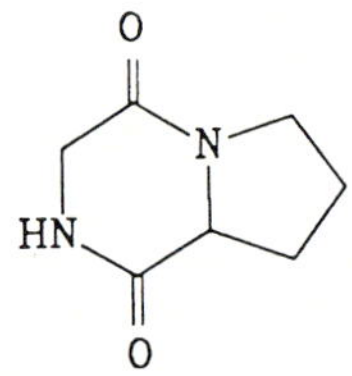

100

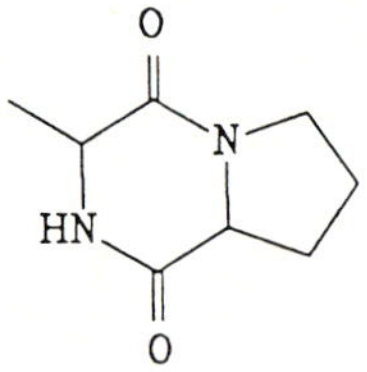

101

reticulatus [105]. The polychlorinated derivatives found in these separate collections of *Dysidea* from Australia notably contrasted with the terpenoid metabolites normally associated with sponges of the Dysideae family. This observation coupled with the fact that blue-green algal symbionts have been reported from these species has led to speculation that these compounds are of algal origin [103].

The only other diketopiperazine derivatives reported from a sponge were isolated from *Tedania ignis* collected in the Caribbean. *Cyclo*-(*S*)-Pro-(*S*)-Leu (**99**), *cyclo*-(*S*)-Pro-(*S*)-Val (**100**), and *cyclo*-(Pro-Ala) (**101**) were identified by spectral analysis and comparison to synthetic samples [110]. Since **99** and **101** had been previously isolated from fungi and plants and were found in only small amounts in *T. ignis*, it is possible that they also are of microbial origin.

The remaining peptide-like compounds isolated from sponges are characterized by the presence of a bromopyrrole moiety. Oroidin (**102**) was isolated from *Agelas oroides* collected in the Bay of Naples [111, 112], the Mediterranean sponges *Axinella damicornis* and *A. Verracosa* [113, 114], the Red Sea sponge *Acanthella aurantiaca* [114], and the New Caledonian sponge *Pseudaxinyassa cantharella* [115]. Oroidin contains a 4,5-dibromopyrrole-2-carboxamide linked by a three carbon chain to a 2-aminoimidazole ring. The structure of oroidin was originally assigned as **103** on the basis of spectral analysis and degradation studies of the natural product and a didehydro derivative [111]. However, synthesis of the

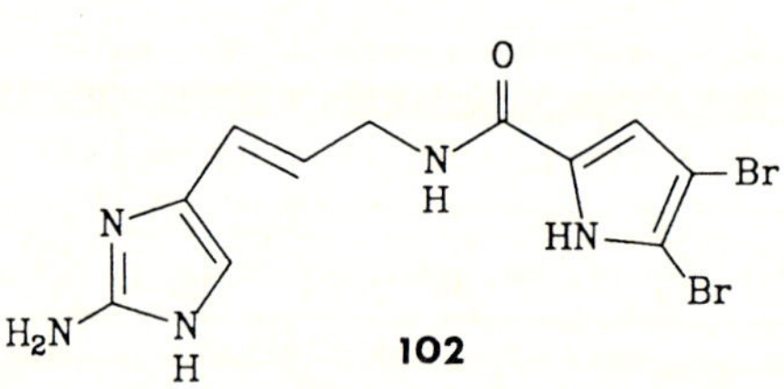

102

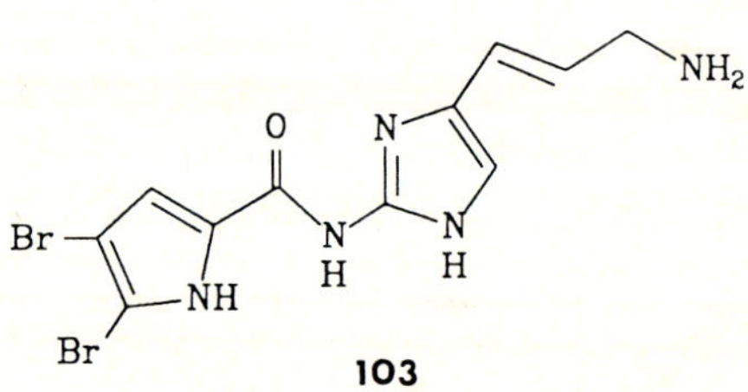

103

imidazole **104** and comparison of the UV data with that of didehydro **105** led to revision of oroidin to structure **102** [112]. Oroidin has been recently synthesized [116]. An Okinawan *Agelas* sp. produces keramadine (**106**), a debromo *cis* isomer of oroidin. Keramadine destroys contractile response to serotonin in rabbit aorta but does not affect responses to either potassium chloride or norepinephrine in the same system [117].

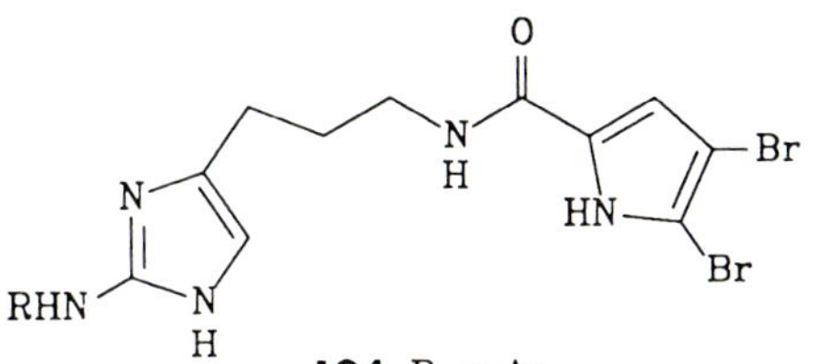

104 R = Ac

105 R = H

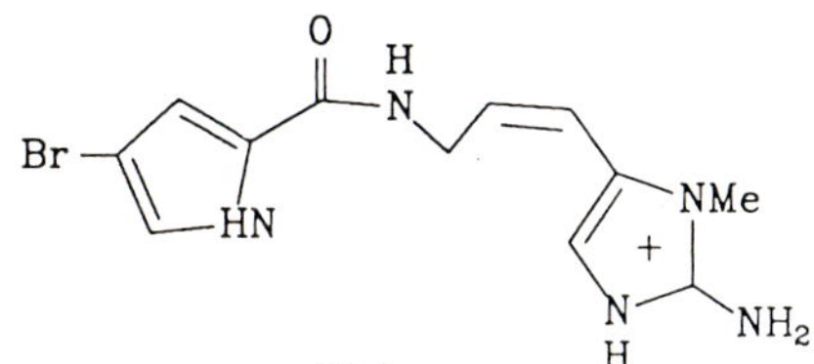

106

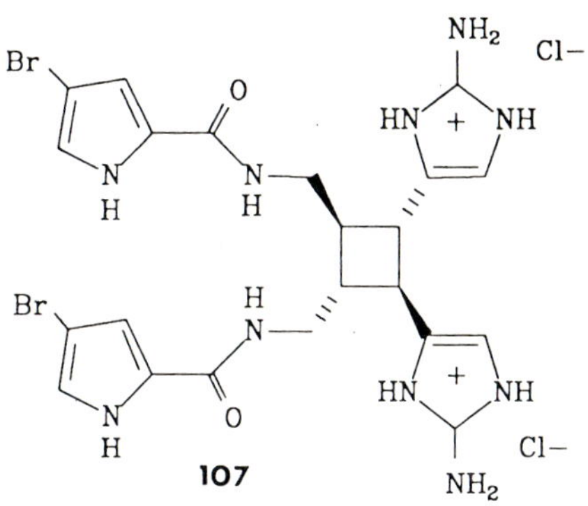

107

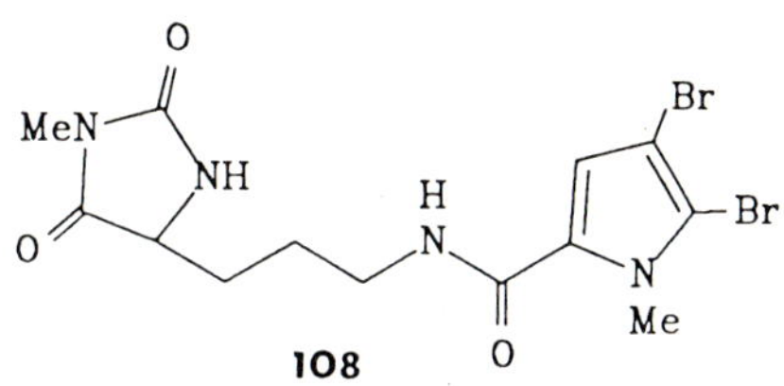

108

Sceptrin (**107**), isolated as the major antimicrobial component of *Agelas sceptrum* (Lamarck) from Belize, is a symmetrical dimer of debromooroidin formed by a head-to-head [2+2] cycloaddition reaction. The stereochemistry and structure of sceptrin were confirmed by X-ray analysis. Sceptrin is more potent than oroidin against *S. aureus* (MIC 15 μg/mL), and also inhibited the growth of *B. subtilis*, *C. albicans*, *P. aeruginosa*, *Alternaria* sp., and *Cladosporium cucumerinum* [118]. The final open chain metabolite in this bromopyrrole series, midpacamide (**108**), was isolated from an unidentified orange sponge collected at Enewetak [119]. Amino acid analysis of midpacamide gave ornithine, and base hydrolysis yielded 1-methyl-4,5-dibromopyrrole-2-carboxylic acid, a second naturally occurring component of the sponge, thus confirming structure **108**.

Several tetracyclic guanidine metabolites related to oroidin (**102**) have also been reported. Dibromophakellin (**109**) and the monobromo analogue **110** were first isolated as their hydrochlorides from the sponge *Phakellia flabellata* collected along the Great Barrier Reef [120, 121]. Both **109** and **110** exhibited very mild antibacterial activity against *B. subtilis* and *E. coli* [121]. The structures were determined by spectral analysis and then confirmed by X-ray diffraction analysis

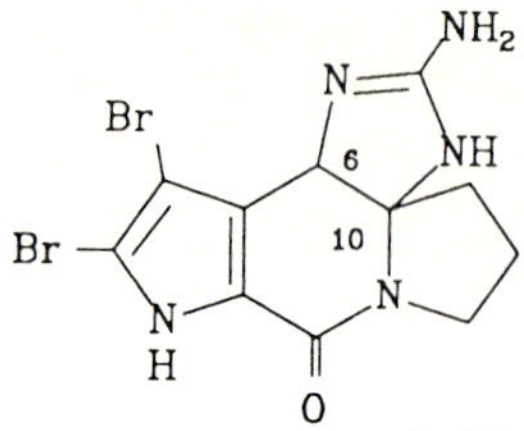

112 9R, 10S

113 9S, 10R

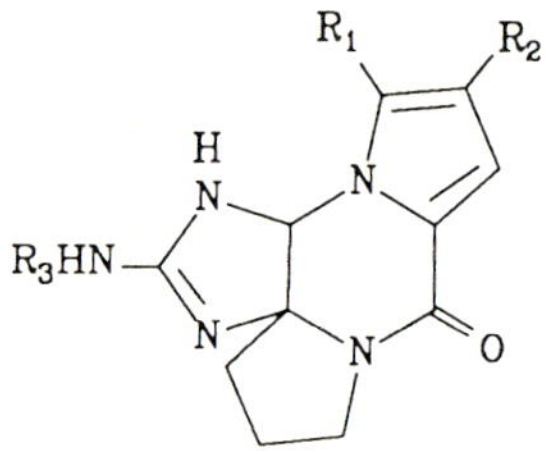

109 $R_1 = R_2 = Br$, $R_3 = H$

110 $R_1 = Br$, $R_2 = R_3 = H$

111 $R_1 = R_2 = Br$, $R_3 = Ac$

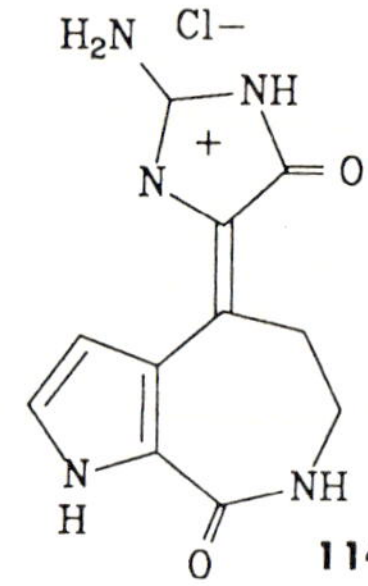

114

on monoacetyl derivative **111**. Recently, the (+) enantiomer of dibromophakellin was found in the New Caledonian sponge *Pseudaxinyssa cantharella* [115], which also produced oroidin and a geometrical isomer of **107**, dibromocantharelline (**112**). Dibromoisophakellin hydrochloride (**113**), the enantiomer of **112**, has been characterized from *Acanthella carteri* collected off the coast of Madagascar [122]. The structures of **112** and **113** were both determined by comparison of spectroscopic data to that of dibromophakellin and ultimately confirmed by X-ray analysis. The phakellins are the only cyclic guanidines containing an endocyclic double bond and may be derived biosynthetically from dihydrooroidin as shown in Scheme 1 [121]. This biogenetic route may cause the guanidinium double bond to be locked in place and may help explain the low basicity demonstrated by this class of compounds.

A series of bi- and tricyclic pyrrololactams related to the phakellins have also been characterized. The "yellow compound" **114** was isolated as its hydrochloride salt first from the Great Barrier Reef sponge *P. flabellata* [123] and later from

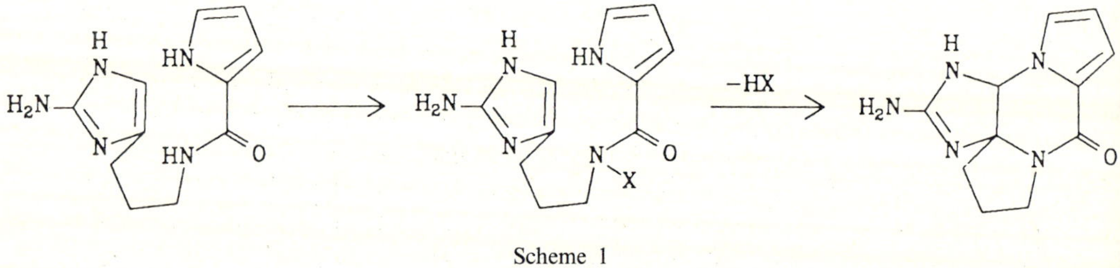

Scheme 1

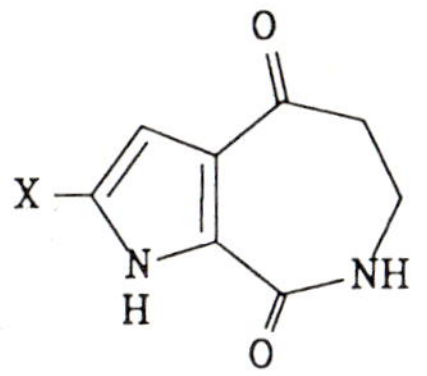

115 X = H

119 X = Br

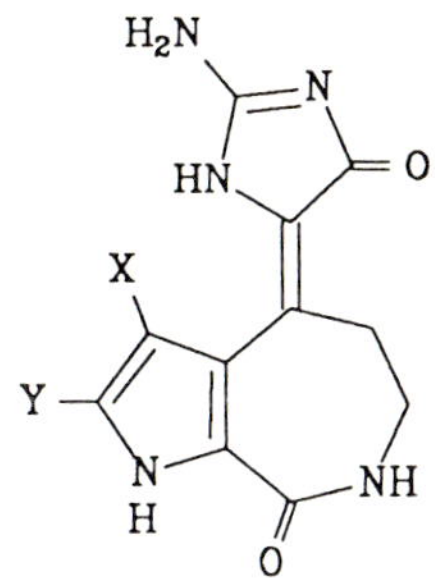

116 X = H, Y = Br

117 X = Y = H

118 X = Y = Br

Hymeniacidon aldis de Laubenfels collected in Okinawa [124]. The structure of **114** was determined by spectroscopic analysis of the natural product and its $KMnO_4$ oxidation product **113** [123]. Compound **115**, for which the name aldisin has been proposed [125], was recently identified as a natural product in extracts of *A. carteri* from Madagascar [126], *H. aldis* from Guam [125], *P. cantharella* from New Caledonia [115], and an unidentified Fijian sponge [125]. The monobromo free base of the "yellow compound" **116**, has been obtained from the Mediterranean sponge *Axinella verrucosa* [14], the Red Sea sponge *Acanthella aurantica* [114],

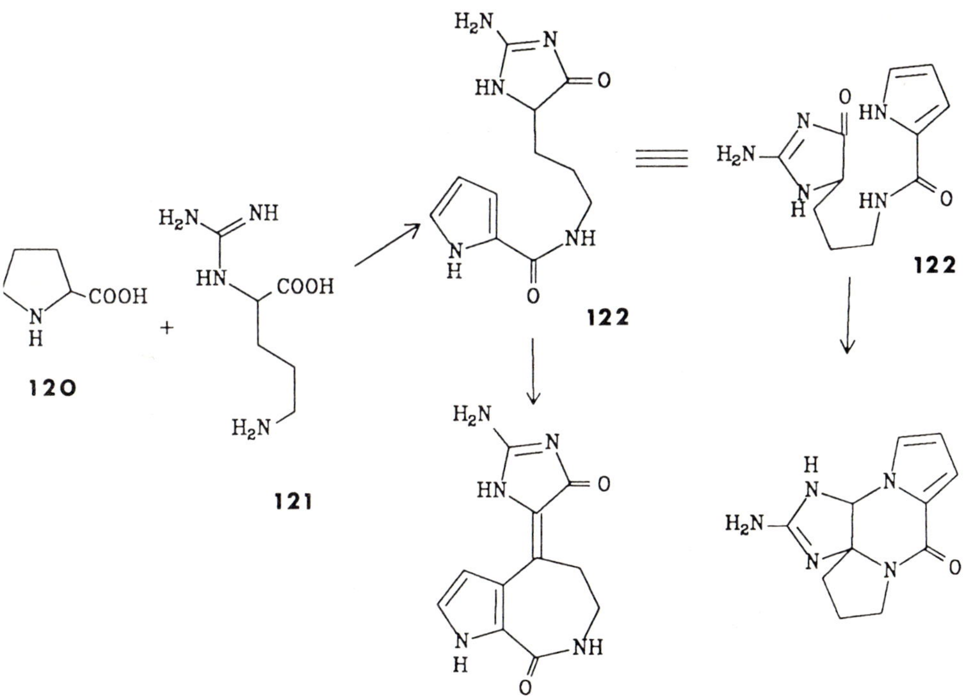

Scheme 2

and also from *H. aldis* (where it was identified as hymenaldisine) [124], *A. carteri* [126], and *P. cantharella* [115]. The structure of **116** was established by spectroscopic and X-ray crystallographic analyses [114, 124]; it demonstrated moderate in vitro cytotoxicity to KB cells but was inactive towards P388 leukemia cells in vivo [114]. The debrominated free base **117** was found in *H. aldis* [124], *A. carteri* [126], and *P. cantharella* [115]. *P. cantharella* also contained the dibrominated metabolite odiline (**118**) [115]. The pyrrololactam 2-bromoaldisin (**119**), which is probably an oxidation product of the tricyclic guanidine derivative **115**, was characterized from extracts of *H. aldis* from Guam [125], a *Lissodendoryx* sp. from Sri Lanka [125], an unidentified Fijian sponge [125], and the New Caledonian sponge *P. cantharella* [115]. Structure **119** was determined by comparison of spectral data to that of other members in this series. The pyrrololactams represent a new class of peptides which may be biosynthesized from proline **120** and the modified α-guanidino isomer of arginine **121** as shown in Scheme 2 [124]. An alternative route could also give the phakellin skeleton and demonstrates the biogenetic relationship between these two groups of metabolites through the oroidin-like intermediate **122**.

4 Tunicates (Ascidians)

When compared to most other marine invertebrates, the tunicates (sea squirts), members of the phylum Chordata, have received relatively little attention from marine chemists. However, within the approximately 50 tunicate metabolites reported are several peptides with significant biological activity and function.

The first peptides isolated from a tunicate were reported in 1980, from *Lissoclinum patella* [127]. There are no terrestrial counterparts to the *Lissoclinum* family of cyclic peptides which now consists of 12 members, all containing at least

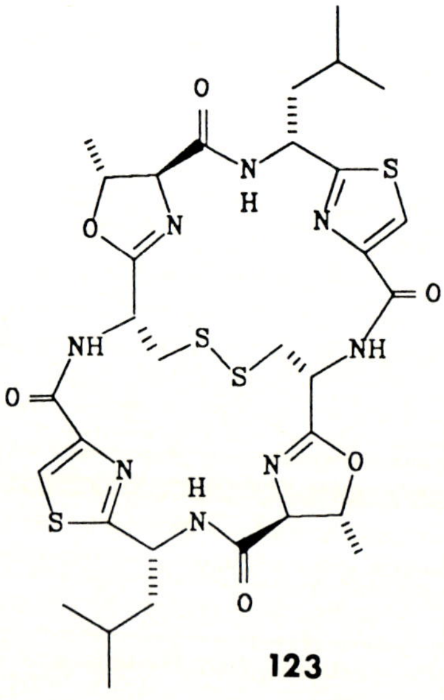

123

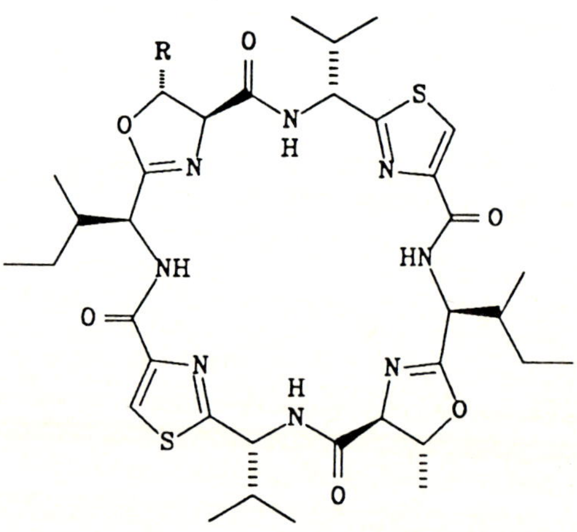

124 R = H

127 R = CH_3

one thiazole and usually an oxazoline amino acid [128, 129, 130, 131]. The *Lissoclinum* peptides can be placed into three subgroups based on structure: (1) ulithiacyclamide (**123**), (2) the patellamide group which includes patellamides A (**124**), B (**125**) and C (**126**) and ascidiacyclamide (**127**), and (3) the lissoclinamide group, which contains lissoclinamides 1 (**128**), 2 (**129**) and 3 (**130**) and ulicyclamide (**131**). All but one member of the family, ascidiacyclamide, were isolated from *L. patella* collected at Palau.

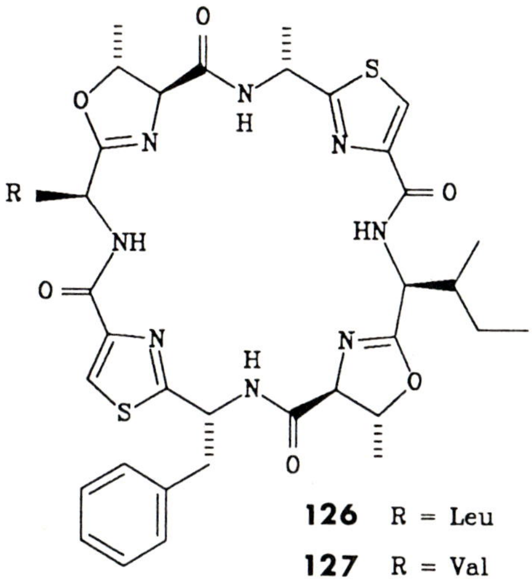

126 R = Leu

127 R = Val

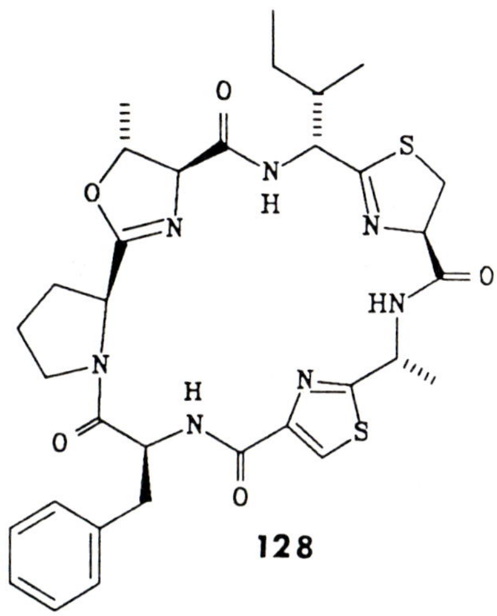

128

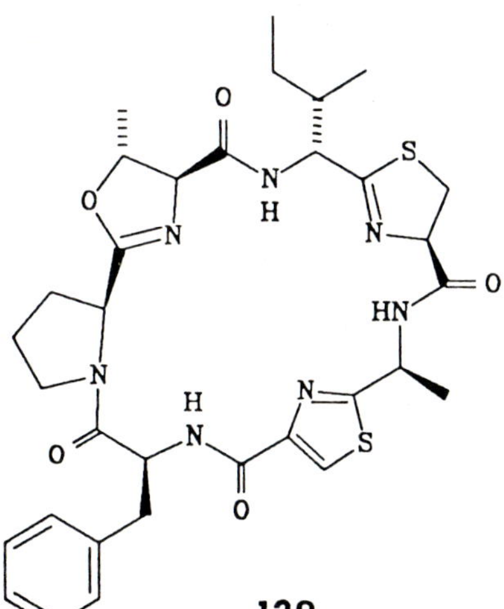

129

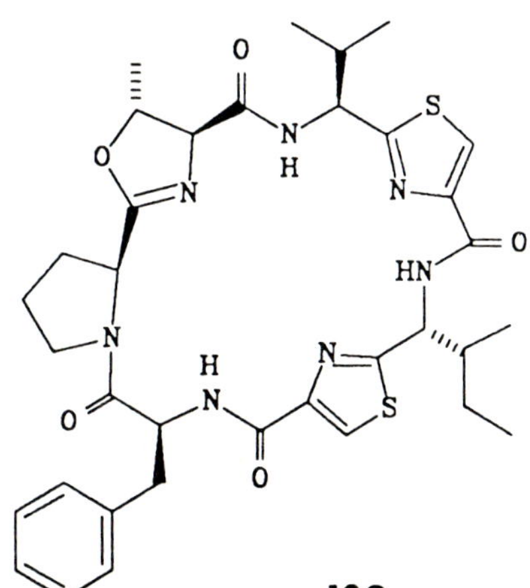

130

Three minor constituents of *L. patella* have been isolated and characterized as prelissoclinamide-2 (**132**), prepatellamide-B-formate (**133**) and preulicyclamide (**134**). These compounds lack an oxazoline present in the namesake peptides and are considered to be biosynthetic intermediates in the peptide pathway rather than

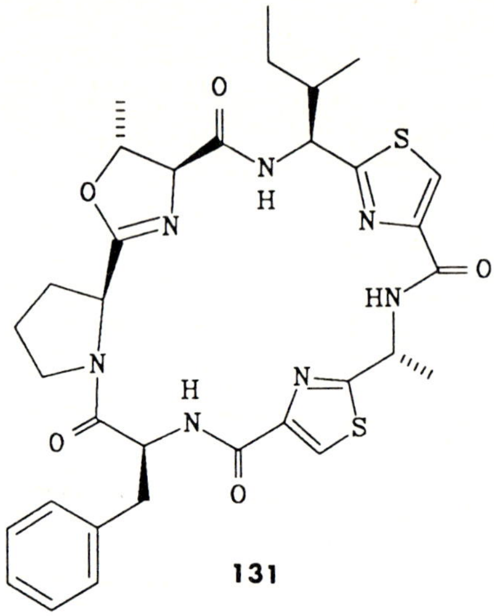

131

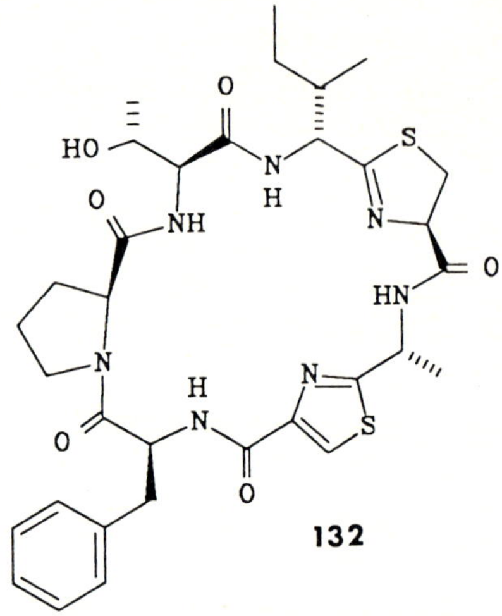

132

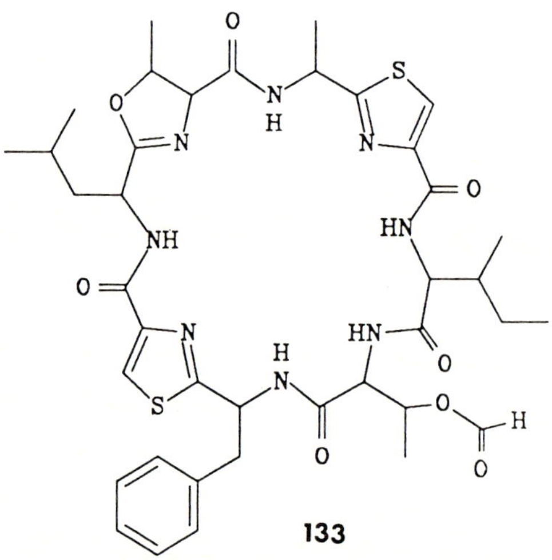

133

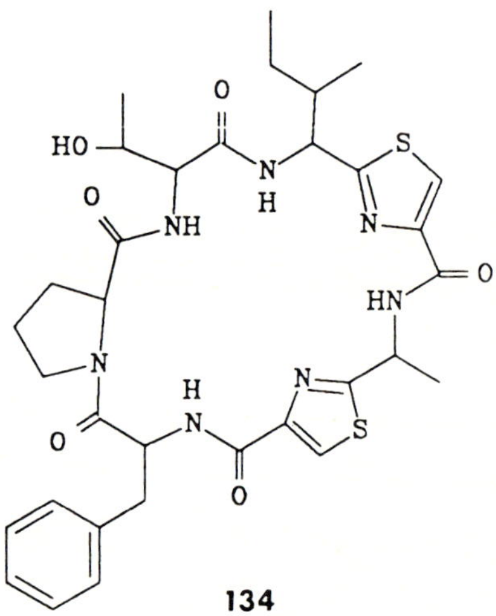

134

degradation products. The conditions for converting lissoclinamide-2 to **132** (5% H_2SO_4 MeOH reflux) are harsher than extraction conditions and no degradation of **129**, **125**, or **131** is observed after extended storage [131].

In the course of solving the structures of these peptides, a new method for establishing the absolute configuration of thiazole amino acids was developed. This method is based on the reaction of thiazoles with singlet oxygen to form a cycloadduct which upon hydrolysis gives an α-amino acid [132]. A large majority of the thiazoles were found to possess the (*R*) absolute configuration. The work on these compounds also resulted in a new method for sequencing small peptides based on the observation of homoallylic coupling between α-protons of α-amino acids using a COSY-45 experiment [131]. This, in part, helped to revise the structures of patellamides A–C (the original assignment placed the thiazoles attached to C-2 of the oxazolines) based on the observation of 5-bond couplings between the α-oxazoline and α-thiazole protons in the COSY-45 spectra. This coupling was not detectable in the conventional one-dimensional spectrum and was determined to be less than 0.2 Hz. The revised structures **124–126** were also confirmed by synthesis [133, 134, 135, 136].

Ulithiacyclamide is the most potent of the *Lissoclinum* peptides, exhibiting in vitro anticancer activity against L1210 (0.1 μg/mL), HeLa (0.1 μg/mL), and CEM (0.01 μg/mL) cell lines [128], and in vivo activity against the P1534J murine leukemia (T/C 188 at 1 mg/kg, repetitive doses) [137]. The oxazoline ring apparently plays an important role in the biological activity of ulithiacyclamide as indicated by a dramatic decrease in cytoxicity for **135** (L1210 $IC_{50} > 125$ μg/mL) [138].

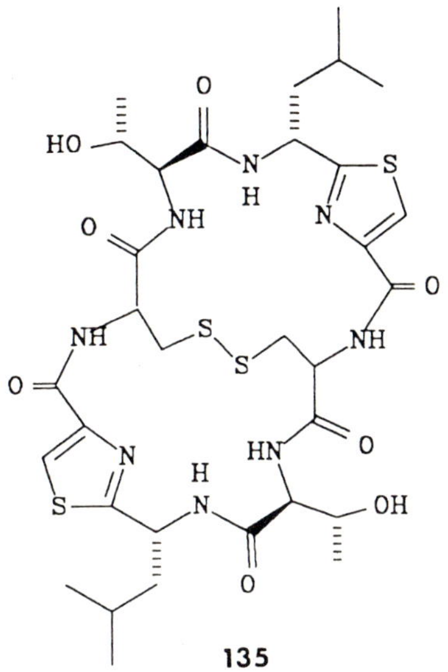

135

The didemnins A–C (**136–138**) are a new class of cyclic depsipeptides isolated from a Caribbean tunicate *Trididemnum* sp. [139, 140]. The didemnins contain a new structural component for depsipeptides, hydroxyisovalerylpropionate (HIP), and were reported to contain a new *allo* stereoisomer of statine. Recently, it was disclosed that didemin B has been synthesized and the structure corrected by replacing statine with isostatine [141]. In addition to their structural novelty, the didemnins also exhibited impressive in vitro and in vivo antiviral activity. Didemnins A and B inhibited Herpes simplex viruses I and II at 1.0 μM and 0.05 μM concentrations, Rift Valley fever virus at 1.37 and 0.04 μg/mL, Venezuelan Equine encephalomyelitis at 0.43 and 0.08 μg/mL, and yellow fever virus at 0.4 and 0.08 μg/mL, respectively. Mice infected with Rift Valley fever showed 90% survival when treated with didemin B at 0.25 mg/kg. There were, however, some drug related deaths at this dose [142]. Didemnin B has also demonstrated in vivo anticancer activity against P388 murine leukemia (T/C 199 at 1.0 mg/kg). [140].

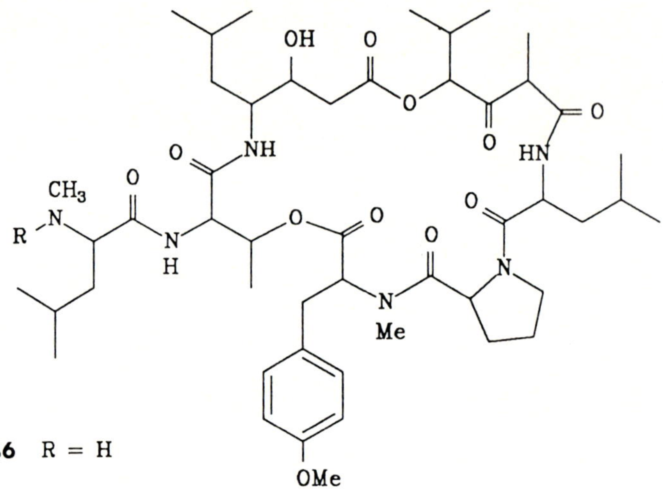

136 R = H

137 R =

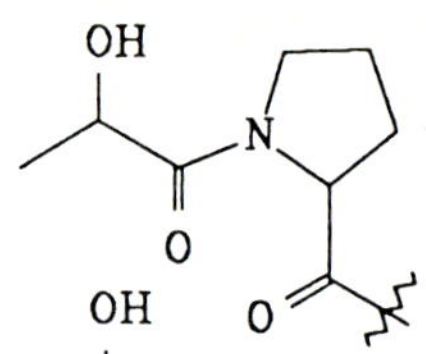

138 R =

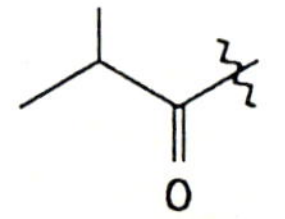

Didemnin B was subsequently evaluated in vitro against human tumors in a stem cell assay [143]. Tumor cells from 8 of 17 patients showed sensitivity to didemnin B with the median IC_{50} being 4.2×10^{-3} μg/mL. Didemnin B is currently in phase II of human clinical trials as an anticancer agent. Added to the impressive list of

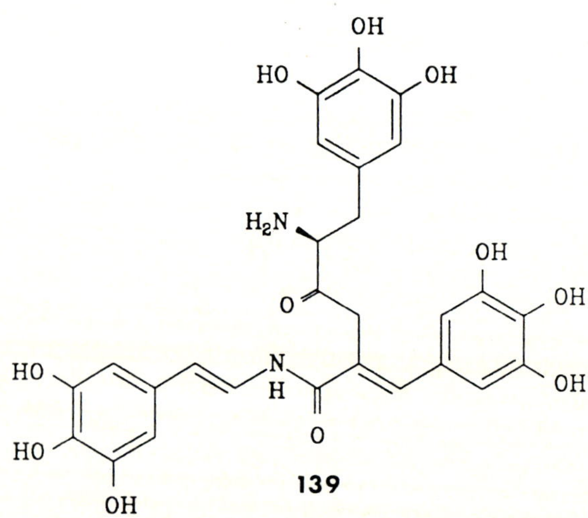

139

activities associated with didemnin B is its effect as an immunosuppressive agent. In a Simonsen parental-to-F_1 graft-versus-host assay, didemnin B showed 71% inhibition of splenomegally with repetitive doses at 0.3 mg/kg [144].

A black Floridian tunicate, *Ascidia nigra*, has been shown to sequester vanadium as the pentavalent vanadate, concentrate it 10^6 fold and store it as the reduced V(III) or V(IV) states at physiological pH. Vanadium (III) at pH greater than 2.5 is stabilized by complexing the metal with a strongly reducing species. The ligands reportedly responsible for this activity are the tunichromes, a series of bright yellow pigments found in the tunicate blood. The structure of one of these pigments, tunichrome B-1 (**139**), was recently established as a modified dopa peptide [145, 146]. The tunichromes, which are extremely sensitive to air and water, were purified by centrifugal counter-current chromatography carried out under dry, O_2-free argon and by using completely degassed solvents. Approximately 6000 tunicates were required to yield 0.5 mg of pure tunichrome B-1.

5 Coelenterates

Coelenterates, which possess nematocysts, or stinging cells that are used to capture prey and for chemical defense, fall into three classes: (1) Anthozoa, the soft corals, gorgonians, and sea anemones, (2) Hydrazoa, which include the Portuguese man-of-war, and (3) Scyphozoa or jellyfish. Peptides, most of which are venoms, have been isolated from all three classes. Many of these are polypeptides and will therefore not be covered as part of this review. A review of this topic is being prepared [147]. Coelenterate toxins can be divided into neuropeptides and neurotoxins.

To date, two neuropeptides from coelenterates have been sequenced. The first of these neuropeptides, a morphogenetic head activator, was isolated from a freshwater hydroid (*Hydra attenuata*) but, as with most neuropeptides, it is present in minute quantities (0.1 pg or 0.1 fmol) [148]. Subsequently, the sea anemone *Anthopleura elegantissima* was found to contain a substance which appeared to be identical by biological and chemical methods to the *Hydra* peptide. Twenty micrograms of peptide were isolated from 200 kg of anemone and the sequence determined as **140** [149]. Structurally, the peptide is unusual in that it lacks a free amino terminus since the terminal glutamate is cyclized to a pyroglutamyl ring. This sequence has been confirmed by synthesis. The head activator is required to initiate head-specific growth and to affect the corresponding cellular differentiation. Since the isolation and sequencing of **140**, head activator-like peptides have also been reported from mammalian hypothalamus and intestine. These mammalian peptides have sequences identical to the hydroid/anemone peptide.

The second group of neuropeptides isolated from coelenterates are related to the molluscan neuropeptide FMRFamide (See Mollusc section). These FMFRamide-like peptides are believed to be ubiquitous in coelenterates, a belief

140 <pEPPGGSKVLF
141 <pEGRG

based on radioimmunoassays using FMRFamide and related RFamide antisera. For example, FMRFamide immunoreactive neurons were found in the ecto- and endoderms of sea anemones (*Metricium senile*), Hydrazoa (*Hydroactine echinata*), soft coral (*Alconium dipitatum*), sea pen (*Pennatula* sp.), and jellyfish (*Pleuroorachia pileus*) [150]. The first FMRFamide-like peptide from a coelenterate, anthro-RFamide (**141**), was isolated from the sea anemone *Anthopleura elegantissima* [151]. The peptide is present in high concentrations in *A. elegantissima* (3.2 nmol/g-wet wt.), much greater than the concentration of the most abundant mammalian neuropeptide present in the brain. The role of **141** is not yet known·but from its probable location in motor neurons, there is speculation that it acts as a transmitter at neuromuscular synapses.

The largest group of peptides isolated from coelenterates are the neurotoxins from sea anemones. These neurotoxins can be placed into two basic categories: long and short. The long neurotoxins are all between 45–50 residues in length and can be further subdivided into two groups based on structural types. Long toxins of type 1 were isolated from *Anemonia sulcata* [152, 153, 154] and *Anthopleura xanthogrammica* [155, 156]. These peptides include ATX I (**142**), II (**143**), V (**144**), and the anthropleurins AP-A (**145**), -B (**146**) and -C (**147**) and all show a great deal of sequence homology, indicated by boxes. All long toxins of type 2 were isolated from anemones of the stichodactylid family [157, 158, 159, 160, 161]. As with type 1 toxins, the type-2 toxins (**148**–**150**) also show a large degree of sequence homology. However, between the two types of long toxins there is only a small amount of sequence homology, primarily in the *N*-terminal region.

		A*																	
142	G	AP	CLC	K	SDGP	NT	RGN	SM	SG	TI	W	V	F	GCPSGW	N	NC	EGRA I	IGYCCK	Q
		I*																	
143	G	VP	CLC	D	SDGP	SV	RGN	TL	SG	II	W	L	A	GCPSGW	H	NC	KKHGPT	IGYCCK	Q
																	P*		
144	G	VP	CLC	D	SDGP	SV	RGN	TL	SG	IL	W	L	A	GCPSGW	H	NC	KKHKGT	IGYCCK	
145	G	VS	CLC	D	SDGP	SV	RGN	TL	SG	TL	W	L Y P	S	GCPSGW	H	NC	KAHGPT	IGYCCK	Q
146	G	VP	CLC	D	SDGP	RP	RGN	TL	SG	TL	W	F Y P	S	GCPSGW	H	NC	KAHGPN	IGYCCK	K
147	G	VP	CLC	D	SDGP	SV	RGN	TL	SG	IL	W	L Y P	A	GCPSGW	H	NC	KAHGPT	IGYCCK	K

148	AA	CKCDD	E	GP	DIRT	A	PL	TG	T	VD	LGS	CN	A	GWEKC	ASY	Y	TII	A	D	CCRKKK
149	GN	CKCDD	E	GP	YVRT	A	PL	TG	Y	VD	LGY	CN	E	GWEKC	ASY	Y	SPI	A	E	CCRKKK
150	AS	CKCDD	D	GP	DVRS	A	TF	TG	T	VD	FWN	CN	E	GWEKC	TAV	Y	TPV	A	S	CCRKKK

* indicates microheterogeneity

The second category of sea anemone neurotoxins are the "short" toxins which have about 30 residues. To date only two peptides of this category have been sequenced [162, 163]. ATX III (**151**), isolated along with ATX I, II, and V from *A. sulcata*, contains 27 residues and three disulfide linkages. Although isolated from the same organism as some of long peptides type 1 it shows only a small amount of

151 RSCCPCYWGGCPWGQNCYPEGCSGPKV-OH

152 AGGKSTCCPAMCKYTAGCPWGQCAHCGCS

homology with them: residues 10–13 in ATX III are homologous with the hydrophobic region 26–28 in ATX II. The second peptide **152** was isolated from *Entacmea quadricolor* (*Parasicyonis actinostoloides*). Although **152** shows little sequence homology with ATX III, it acts at the same site on the sodium channel [163].

Despite the fact that there are three types of sea anemone toxins which share little sequence homolgy, all of these toxins have the same mechanism of action. All alter the function of sodium channels in excitable membranes of muscles and nerves primarily by delaying the inactivation process. Because of this, the toxins are considered important probes of Na channels and related physiological functions. These peptides have been studied extensively for their physiological function and mechanism of action [164, 165, 166, 167, 168, 169].

While some of the sea anemone toxin sequences have been known for more than 10 years, their secondary structure in solution has not yet been defined. Through the years, many research groups have tried to define secondary structure by a variety of methods. In 1979, Norton and coworkers used the Chou–Fasman rules to predict the secondary structure of AP-A [170]. Based on their calculations, AP-A was predicted to consist of a series of γ-reverse turns plus a few short regions of β-sheets. This picture of AP-A was apparently supported by laser Raman and circular dichroism spectroscopic studies they had previously undertaken. In contrast, Nabuillin, using the modified Chou–Fasman procedure of Dufton and Hider, proposed that AP-A contains β-sheet, β-turns and 2 segments of helix [171].

Within the last five years, a combination of increased accessibility of higher field NMR (500 MHz) instruments and the development of sophisticated 1D and 2D pulse sequences has led to reinvestigation of the solution conformation of these peptides. The majority of investigators have focused on type-1 toxins, although some work has been done on type-2 peptides as well. Gooley and Norton studied AP-A and ATX I by ^{1}H NMR using nOe, 2D NMR, NH exchange rates and NH-H spin–spin coupling constants [172]. From these experiments it was proposed that AP-A and ATX I contain a region of four short strands of anti-parallel β-sheet with little or no helix. This region of β-sheet brings the aromatic rings of Trp-23 and -33 into close proximity to form the nucleus for a small hydrophobic region. This conformational picture of ATX I and AP-A suggests that neither of the Chou–Fasman predictions is correct. This may be partly due to the presence of three disulfide bonds in a peptide that is only 49 residues long. Overall the data indicates that the structure of AP-A and ATX I are similar in solution. A major difference between the two is that AP-A displays structural heterogeneity (several distinct conformations by NMR) not seen in ATX I. The lack of structural heterogeneity in ATX I is attributed to the replacement of Gly at residue forty-one in ATX I with a Pro in AP-A. Proline shows a marked ability to cause adjacent peptide bonds to adopt a *cis* conformation exclusively and consequently AP-A exists as several distinct conformations in NMR spectra. ATX I, which contains a

Gly at position 41, does not exist in distinct conformations because glycine does not restrict the rotation of its adjacent peptide bonds [170].

6 Molluscs

Marine molluscs produce a large number of amino acid metabolites. Many are used as venoms [173], while others have been implicated in the regulation of sexual and digestive functions [174, 175]. With a few exceptions, these can be classified as unmodified linear peptides comprised of (*S*)-amino acids.

The dolastatins, isolated by Pettit, represent an exception to this rule. This family of cyclic peptides was isolated from the Indian Ocean sea hare *Dolabella auricularia* and first reported in 1982. The dolastatins are powerful cell growth inhibitors of the P388 lymphocytic leukemia cell line with dolastatin 3 (**153**) having a reported ED_{50} of 1×10^{-4} to 1×10^{-7} [176].

Structure **153** was proposed for dolastatin 3 largely from the results of NMR studies and acid hydrolysis experiments which established the amino acid composition, and electron impact mass spectrometry which determined amino acid sequence [176]. The structure of dolastatin 3 became suspect when Shioiri and Schmidt published independent syntheses of all 16 possible stereo and sequence isomers of **153**, none of which possessed the spectroscopic and chromatographic properties of the natural product [177, 178]. Subsequently, Pettit also reported synthesis of structure **153** and its dissimilarity with the natural product [179].

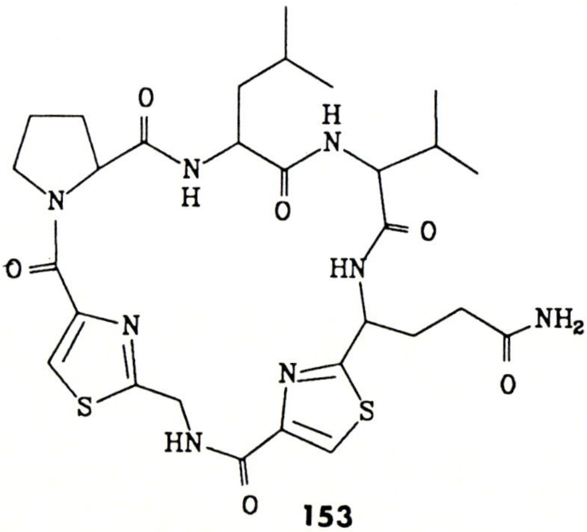
153

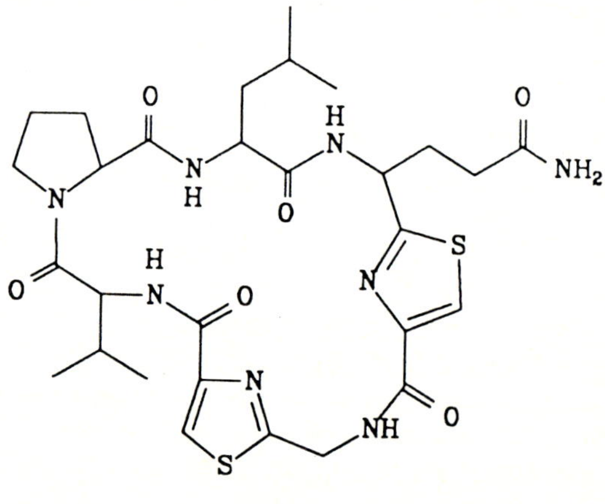
154

Although this synthetic work failed to yield the natural product, the structure of which is still a mystery, it yielded significant advances in the field of peptide synthesis, specifically, the use of a pentafluorophenyl ester ring closure reaction to effect 90% or greater yield of macrocyclic peptide systems [178]. Late in 1987 Pettit published a revision of the dolastatin 3 structure to **154** based on total synthesis

[180]. The revised structure contains the same amino acid components (all with (*S*) absolute configuration) as previously reported. Although the synthetic material is reported to be identical to the natural product, it is 4 to 5 orders of magnitude less cytotoxic than originally reported for dolastatin 3. Also in 1987, Pettit reported the structure of dolastatins 10 (**155**) and 11 (**156**). Dolastatin 10, which contains a number of unusual amino acids including a modified proline and *O*-methyl isostatine is claimed to be the most potent antineoplastic agent known (17–67% curative response to the NCI human melanoma xenograph in mouse at 3.25 μg/kg; T/C up to 238 at 11 μg/kg against B16 melanoma and 202 at 4 μg/kg against P388) [181]. Dolastatin 11 reportedly differs from the cyanophyaceal depsipeptide, majusculamide C (**18**), only in replacement of *N*-methylisoleucine with *N*-methylleucine [182].

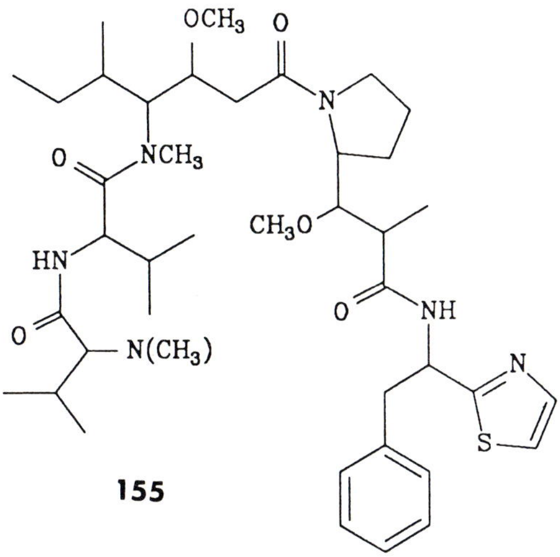

155

156 R = *i*-butyl

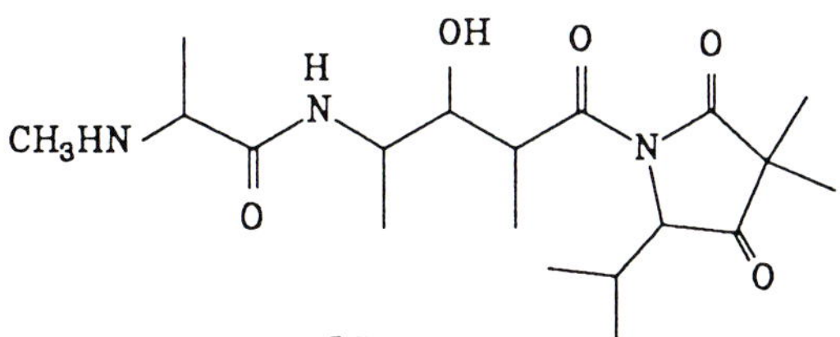

157

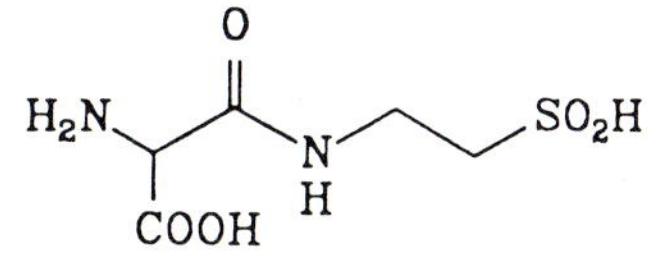

158

Janolusimide (**157**), isolated from *Janolus cristatus*, is the only peptide known from a nudibranch. Janolusimide appears to be derived from condensation of *N*-methyl-(*S*)-alanine, 4-amino-3-hydroxyl-2-methyl valeric acid, valine, and isobutyric acid. Janolusimide is toxic to mice at 5 mg/kg, when administered intraperitoneally, apparently acting as a cholinergic agent. At lower doses its action is antagonized by atropine [183].

Arcamine (**158**), a dipeptide of hypotaurine and 2-aminomalonic acid was isolated from the mussel *Arca zebra* as a fish attractant [184]. At concentrations of 10^{-8} g/L arcamine induced feeding behavior in unspecified aquarium fish.

Opioid-type peptides have also been reported from molluscs. Met- and Leu-enkephalin and Met-enkephalin-Arg-Phe were isolated from neural tissue of the mussel *Mytilus edulis* [185]. These peptides are present at concentrations of 2–3.5 ng per ganglion.

FMRFamide (**159**) and a number of related peptides were originally isolated from cerebral, pedal, and visceral ganglia of the venus clam *Macrocallista nimbosa* [186]. Subsequently, several analogs of MRFamide including a family of heptapeptides (e.g., **160**) have been reported from molluscs [187]. A wide range of activities on cardiac and noncardiac muscle tissue from several mollusc species have been reported, and at least three types of receptors have been defined [188, 189]. Much of the work in this area concerns the pharmacology rather than chemistry of these peptides and has recently been reviewed [190, 191].

159 Phe-Met-Arg-Phe-NH_2

160 p-Glu-Asp-Pro-Phe-Leu-Arg-Phe-NH_2

161 H-ISINQDLKAITDMLLTEQIQARRRCLDALRQRLLDL-NH_2
H-DSDVSLFNGDLLPNGRCS-OH

162 H-ISINQDLKAITDMLLTEQIQARQRCLAALRQRLLDL-NH_2
H-DSDVSLFNGDLLPNGRCS-OH

163 H-ISINQDLKAITDMLLTEQIQARRRCLAALRQRLLDL-NH_2
H-DSDVSLFNGDLLPNGRCS-OH

164 ISWPSYPGSEGIRSSNCQKKLNCGTKNIATKGVCKAFCLGRKRFWQK-
CGKNCSGSKGSKVCNAVLAHAVEKAGKGLIAVTDKAVAAIVKLAAGIA

The califins (**161**–**163**) isolated from *Aplysia californica* contain 54 amino acid residues each in two subunits linked by a disulfide bond. They possess a high degree of homology differing at only two sites, 23 and 27 in the large subunit. Although the exact physiological role of the califins is unknown, it has been speculated that they act as pheromones [175].

7 Heteronemertine Worms

Few peptides have been isolated from marine worms. Kem and Blumenthal have described the isolation, characterization, sequences, and biological activity of two classes of polypeptides obtained from the mucus secreted from the skin of the heteronemertine worm, *Cerebratulus lacteus* [192, 193]. Toxin A-III (**164**) is a basic polypeptide composed of 95 amino acid residues with disulfide cross linkages between residues 17 and 38, 23 and 34, and 48 and 61. The cytolytic toxin A-III causes rapid lysis of human erythrocytes, Ehrlich Ascites cells and cardiac muscle

165 ASSTWGGSYXACENNCRKQYDDCIKCQGKWAGKRGKCAAHCAVQ-TTSCNDKCKKH
166 ASATWGAAYXACENNCRKKYDLCINCQGKWAGKNGKCAAHCIIQ-KNNCKGKCKKE

cells at concentrations between 1.1 to 3.1 μg/mL. Evidence suggests that penetration of the membrane by the toxin may be important for cytolytic activity [194]. In contrast to toxin A-III, the smaller toxins B-II (**165**), and B-IV (**166**) are neurotoxins which act specifically on the axons of crustaceans by delaying inactivation of sodium channels [195, 196, 197]. Although there is a high degree of homology between **165** and **166**, neither is significantly homologous with scorpion or sea anemone neurotoxins which display similar activity.

8 Vertebrates

Soles (flatfish) of the genus *Pardichirus* have been shown to secrete an ichthyotoxic material from glands lining the dorsal and anal fins [198]. This secretion from *P. marmoratus* (Moses sole) in the Red Sea has demonstrated shark-repellent activity [199] and the secretion of *P. pavoninus* (Peacock sole) in the western Pacific presumably does the same [198]. The toxicity and shark-repellent activity of the secretion has been shown to belong, in part, to groups of saponins [200, 201]. However, in 1978, Primor reported the isolation and partial structure of an ichthyotoxic agent named pardaxin from the defense secretion of *P. marmoratus* [202]. In 1986 Thompson reported the isolation and primary structure of three toxic peptides from *P. pavoninus* called pardaxins P-1 (**167**), -2 (**168**) and -3 (**169**) [203]. These new peptides are believed to be the components of the substance originally described as pardaxin, based on comparison with an authentic sample from *P. marmoratus.*

167 GFFALIPKIISSPLFKTLLSAVGSALSSSGEQE
168 GFFALIPKIISSPIFKTLLSAVGSALSSSGGQE
169 GFFAFIPKIISSPLFKTLLSAVGSALSSS(GG)QE

Both pardaxin P-1 and P-2 cause death in *Oryzias latipes* (killi-fish) in 30 minutes at a concentration of 25 μg/mL. Paradaxin P-3 was not isolated in large enough quantities to allow proper bioassays. An ichthyotoxic, saponin-free fraction of the crude secretion was also shown to possess shark-repellent activity against the white-tip reef shark *Teiaenodon obesus.* A clear escape response was demonstrated when 100 mg of this peptide fraction was dissolved in five mL of seawater and injected into the shark's mouth.

The pardaxins also exhibit pharmacological similarities to the bee venom melittin, including lysis of erythrocytes, strong surfactant activity, and also require the amino termini for activity [204].

Tropical marine fish from several families have been shown to secrete toxic peptides from specialized glands in the skin. The peptides known as grammistins appear to be universal in the secretions of the Grammistidae (soapfishes) [205, 206,

207, 208]. Similar compounds have also been isolated from species of *Gobiodon* and the clingfish *Diademichthys lineatus* [209, 210]. All of these substances exhibit both ichthyotoxic and hemolytic activity and impart a bitter taste to the mucus of the skin; therefore, they are believed to function at least in part as a defense against predators. While no primary structures have been reported for any of these peptides the amino acid compositions of some are known [210, 211].

9 References

1. Bergmann W, Feeney RJ (1980) J. Am. Chem. Soc. 72:2809
2. Ireland CM, Roll DM, Molinski TF, McKee TC, Zabriskie TM, Swersey JC (1987) Proc. Calif. Acad. Sci. 13:41–58
3. Faulkner DJ (1977) Tetrahedron 33:1421
4. Scheuer PJ (1978) In: Marine Natural Products: Chemical and Biological Perspectives, Vol I, Academic, New York
5. Fenical W (1975) J. Phycol. 2:245
6. Fenical, W (1986) Marine alkaloids and related compounds. In: Pelletier, SW (ed) Alkaloids: Chemical and Biological Perspectives, Vol IV, John Wiley, New York, p 275
7. Chevolet L (1981) Guanidine derivatives. In: Scheuer PJ (ed) Marine Natural Products: Chemical and Biological Perspectives, Vol IV, Academic, New York, chap 2
8. Fattorusso E, Piattelli M (1980) Amino acids from marine algae. In: Scheuer PJ (ed) Marine Natural Products: Chemical and Biological Perspectives, Vol III, Academic, New York, chap 2
9. Miyazawa K (1971) Nippon Suisan Gakkaishi 37:788
10. Shiba T, Wakamiya T (1984) Kagaku (Kyoto) 39:415; Chem Abs. 100:153822c
11. Faulkner DJ (1984) Nat. Prod. Rep. 1:251
12. Faulkner DJ (1985) Nat. Prod. Rep. 1:552
13. Faulkner DJ (1986) Nat. Prod. Rep. 3:1
14. Krebs HC (1986) Prog. Chem. Org. Nat. Prod. 49:152
15. Wakamiya T, Seshima T, Shiba T (1986) Iyo Masu Kenkyukai Koenshu 11:35; Chem. Abs. 106:134466x
16 Wakamiya T, Shiba T (1987) Kagaku Zokan (Kyoto): 21; Chem. Abs. 107:15116r
17. Lewis EJ (1962) Curr. Sci. 31:90
18. Lewis EJ, Gonzalves EA (1962) Ann. Bot. (London) 26:301
19. Sciuto S, Piatelli M, Impellizzeri G (1978) Phytochem. 17:1659
20. Miyazawa K, Ito K (1974) Nippon Suisan Gakkaishi 40:815; Chem. Abs. 81:14859u
21. Nakamoto H, Wakamiya T, Shiba T, Matsumiya, Setogawa K (1984) Abstracts of the 49th Japanese Chemical Society Conference, Abst. 3(2)
22. Matsutani H, Setogawa K, Wakamiya T, Nakamoto H, Shiba T (1987) Nippon Suisan Gakkaishi 53:509
23. Ohira T (1939) J Agric. Chem. Soc. Jpn. 15:370; Chem. Abs. 33:6254
24. Konayaga S (1966) Nippon Suisan Gakkaishi 32:967
25. Ohira T (1940) J. Agric. Chem. Soc. Jpn. 16:1; Chem. Abs. 34:3683
26. Ohira T (1940) J. Agric. Chem. Soc. Jpn. 16:293: (1951) Chem. Abs. 45:4753
27. Ohira T, Kido H, Imai K (1955) Nippon Nogei, Kagaku Kaishi 29:520; (1959) Chem. Abs. 53:1169
28. Tazawaka Y (1949) Act Phytochim. 15:141; (1950) Chem. Abs. 44:3441
29. Kaneko T, Shiba T, Wataria S, Imai S, Shimada T, Ueno K (1957) Chem. and Ind. (London) 986

30. Rudinger J, Pravda Z (1958) Chem. Listy 52:120
31. Shiba T, Imai S (1959) Nippon Kagaku Zasshi 80:492; (1961) Chem. Abs. 55:3452
32. Dekker CA, Stone D, Fruton JS (1949) J. Biol. Chem. 181:719
33. Tagaki M, Iida A, Murayama H, Soma S (1973) Nipon Suisan Gakkaishi 39:961
34. Makisumi S (1959) J. Biochem. (Tokyo) 46:63.
35. Miyazawa K, Ito K, Matsumoto F (1976) Hiroshima Daigaku Suichikusan Gakubu Kiyo 15:161; Chem. Abs. 86:167528
36. Gorham PR, Carmichael WW (1979) Pure Appl. Chem. 52:165
37. Santikarn S, Williams DH, Smith RJ, Hammond SJ, Botes DP, Tuiman AA, Wessels PA, Viljoen CC, Kruger HA (1983) J. Chem. Soc., Chem. Commun. 652
38. Meeting B, Pyne JW (1986) Enzyme Microb. Technol. 8:386
39. Sivonen K, Himberg K, Niemla S (1986) Kem-Kemi 13:396; Chem. Abs. 105:48492a
40. Ueno Y (1986) Pure Appl. Chem. 58:39
41. Jackson ARB, Runnegar MTC, Falconer IR, McInnes A (1985) Cyanobacterial (blue-green algae) Toxicity of Livestock, In: Seawright AA (ed) Plant Toxicology, Proceedings of Australia-USA Poisonous Plants Symposium, Yeerongpilly, Australia, p 499
42. Codd GA (1984) Microbiol. Sci. 1:48
43. Botes DP, Tuiman AA, Wessels PL, Viljoen CC, Kruger H, Williams DH, Sanatikarn S, Smith RJ, Hammond SJ (1984) J. Chem. Soc., Perkin Trans. I 2311
44. Botes DP, Wessels PL, Kruger H, Runnegar MTC, Santikarn S, Smith RJ, Barna JCJ, Williams DH (1985) J. Chem. Soc., Perkin Trans. I 2747
45. Elleman TC, Falconer IR, Jackson ARB, Runnegar MTC (1983) Aust. J. Biol. Sci. 31:209
46. Runnegar MTC, Falconer IR (1982) S. Afr. J. Sci. 78:363
47. Kusumi T, Ooi T, Watanabe M, Kakisawa H (1987) Tennen Yuki Kagobutsu Toronkai Koen Yoshishu 29:536; Chem. Abs. 108:17524v
48. Kusumi Y, Ooi T, Watanabe M, Takahashi H, Kakisawa H (1987) Tetrahedron Lett. 28:4695
49. Hashimoto Y (1979) Marine toxins and other bioactive marine metabolites. Japan Scientific Societies Press, Tokyo
50. Moikeha SN, Chu GW, Berger LB (1971) J. Phycol. 7:4
51. Hashimoto Y, Kamiya H, Yamazato K, Nozawa K (1976) In: Ohsaka A, Hayashi K, Sawai Y (eds) Animal, plant and microbial toxins, Vol. 1, Plenum, New York, p 333
52. Cardellina II JH, Marner FJ, Moore RE (1979) Science 204:193
53. Takashima M, Sakai H (1960) Bull. Agri. Chem. Soc. Jpn. 24:647
54. Sakai S, Hitotsuyanagi Y, Aimi N, Fujiki H, Suganuma M, Sugimura T, Endo Y, Shudo K (1986) Tetrahedron Lett. 27:5219
55. Fujiki H, Sugimura T (1983) Cancer Surveys 2:539
56. Fujiki H, Suganuma M, Tahira T, Yoshioka A, Nakayasu M, Endo Y, Shudo K, Takayama S, Moore RE, Sugimura T (1984) New Classes of Tumor Promoters: Teleocidin, Plysiatoxin, and Palytoxin. In: Fujiki H, (ed) Cellular interactions by environmental tumor promoters. Japan Scientific Societies Press, Tokyo/VNU Science Press, Utrecht, p 37
57. Fujiki H, Mori M, Nakayasu M, Terada M, Sugimura T, Moore RE (1981) Proc. Natl. Acad. Sci. USA 78:3872
58. Muratake H, Natsume M (1987) Tetrahedron Lett. 28:2265
59. Marner FJ, Moore RE, Hirotsu K, Clardy J (1977) J. Org. Chem. 42:2815
60. Carter DC, Moore RE, Mynderse JS, Niemczura WP, Todd JS (1984) J. Org. Chem. 49:236
61. Gregson JM (1986) Isolation and structure determination of the puwainaphycins A-D M. Sc. Thesis, University of Hawaii, Honolulu

62. Bax A, Lerner L (1986) Science 232:960
63. Matsunaga S, Fusetani N, Konosu S (1985) J. Nat. Prod. 48:236
64. Matsunaga S, Fusetani N, Konosu S (1984) Tetrahedron Lett. 25:5165
65. Matsunaga S, Fusetani N, Konosu S (1985) Tetrahedron Lett. 26:855
66. Kitagawa I, Kobayashi M, Lee NK, Shibuya H, Kawata Y, Sakiyama F (1986) Chem. Pharm. Bull. 34:2664
67. Nakamura H, Kobayashi J, Nakamura Y, Ohizumi Y, Kondo T, Hirata Y (1986) Tetrahedron Lett. 27:4319
68. Kitagawa I, Lee NK, Kobayashi M, Shibuya H (1987) Chem. Pharm. Bull. 35:2129
69. Zabriskie TM, Klocke JA, Ireland CM, Marcus AH, Molinski TF, Faulkner DJ, Xu C, Clardy JC (1986) J. Am. Chem. Soc. 108:3123
70. Crews P, Manes LV, Boehler M (1986) Tetrahedron Lett. 27:2797
71. Braekman JC, Daloze D, Moussiaux B, Riccio R (1987) J. Nat. Prod. 50:994
72. Chan WR, Tinto WF, Manchand PS, Todaro LJ (1987) J. Org. Chem. 52:3091
73. Quinoa E, Adamczeski M, Crews P, Bakus GJ (1986) J. Org. Chem. 51:4497
74. Andersen RJ (1978) Tetrahedron Lett. 2541
75. Andersen RJ, Stonard RJ (1979) Can. J. Chem. 57:2325
76. Schmidt U, Lieberknecht A, Griesser H, Bokens H (1985) Liebigs Ann. Chem. 785
77. Stonard RJ, Andersen RJ (1980) J. Org. Chem. 45:3687
78. Stonard RJ, Andersen RJ (1980) Can. J. Chem. 58:2121
79. Schmidt U, Wild J (1984) Angew. Chem. Int. Ed. Engl. 23:991
80. Kazlauskas R, Lidgard RO, Murphy PT, Wells RJ (1980) Tetrahedron Lett. 21:2277
81. Kazlauskas R, Lidgard RO, Murphy PT, Wells RJ, Blount JF (1981) Aust. J. Chem. 34:765
82. Nishiyama S, Yamamura S (1982) Tetrahedron Lett. 23:1281
83. Nishiyama S, Suzuki T, Yamamura S (1982) Tetrahedron Lett. 23:3699
84. Fattorusso E, Minale L, Sodano G, Moody K, Thomson RH (1970) J. Chem. Soc., Chem. Commun. 752
85. Moody K, Thomson RH, Fattorusso E, Minale L, Sodano G (1972) J. Chem. Soc., Perkin Trans. I 18
86. Cimino G, De Rosa S, De Stefano S, Self R, Sodano G (1983) Tetrahedron Lett. 24:3029
87. McMillan JA, Paul IC, Goo YM, Rinehart Jr. KL, Krueger WC, Pschigoda LM (1981) Tetrahedron Lett. 22:39
88. Nishiyama S, Yamamura S (1983) Tetrahedron Lett. 24:3351
89. Thompson JE, Barrow KD, Faulkner DJ (1983) Acta Zoologica 64:199
90. Arabshahi L, Schmitz FJ (1987) J. Org. Chem. 52:3584
91. Quinoa E, Crews P (1987) Tetrahedron Lett. 28:3229
92. Rodriguez AD, Akee RK, Scheuer PJ (1987) Tetrahedron Lett. 28:4989
93. Gopichand Y, Schmitz FJ (1979) Tetrahedron Lett. 3921
94. Borders DB, Morton GD, Wetzel EK (1974) Tetrahedron Lett. 2709
95. Nakamura H, Wu H, Kobayashi J, Nakamura Y, Ohizumi Y, Hirata Y (1985) Tetrahedron Lett. 26:4517
96. Litaudon M, Guyot M (1986) Tetrahedron Lett. 27:4455
97. Kashman Y, Groweiss A, Carmely S, Kinamoni Z, Czarkie D, Rotem M (1982) Pure Appl. Chem. 54:1995
98. Rotem M, Carmely S, Kashman Y, Loya Y (1983) Tetrahedron 39:667
99. Roll DM, Chang CWJ, Scheuer PJ, Gray GA, Shoolery JN, Matsumoto GK, Van Duyne GD, Clardy J (1985) J. Am. Chem. Soc. 107:29165

100. Bergquist PR, Wells RJ (1983) Chemotaxonomy of the Porifera: the development and current status of the field. In: Scheuer PJ (ed) Marine Natural Products: Chemical and Biological Perspectives, Vol V, Academic, New York, chap 1
101. Hofheinz W, Oberhansli WE (1977) Helv. Chim. Acta 60:660
102. Williard PG, de Laszlo SE (1984) J. Org. Chem. 49:3489
103. Kazlauskas R, Murphy PT, Wells RJ (1978) Tetrahedron Lett 4945
104. Kazlauskas R, Lidgard RO, Wells RJ, Vetter W (1977) Tetrahedron Lett. 3183
105. Charles C, Braekman JC, Daloze D, Tursch B, Karlsson R (1978) Tetrahedron Lett. 1519
106. Charles C, Braekman JC, Daloze D, Tursch B (1980) Tetrahedron 36:2133
107. Erickson KL, Wells RJ (1982) Aust. J. Chem. 35:31
108. Biskupiak JE, Ireland CM (1984) Tetrahedron Lett. 25:2935
109. de Laszlo SE, Williard PG (1985) J. Am. Chem. Soc. 107:199
110. Schmitz FJ, Vanderah DJ, Hollenbeak KH, Enwall CEL, Gopichand Y, SenGupta PK, Hossain MB, van der Helm D (1983) J. Org. Chem. 48:3941
111. Forenza S, Minale L, Riccio R, Fattorusso E (1971) J. Chem. Soc., Chem. Commun. 1129
112. Garcia EE, Benjamin LE, Fryer RI (1973) J. Chem. Soc., Chem. Commun. 78
113. Cimino G, De Stefano S, Minale L, Sodano G (1975) Comp. Biochem. Physiol. 50B:279
114. Cimino G, De Rosa S, De Stefano S, Mazzarella L, Puliti R, Sodano G (1982) Tetrahedron Lett. 23:767
115. De Nanteuil G, Ahond A, Guilhem J, Poupat C, Tran Huu Dau E, Potier P, Pusset M, Pusset J, Laboute P (1985) Tetrahedron 41:6019
116. De Nanteuil G, Ahond A, Poupat C, Thoison O, Potier P (1986) Bull. Soc. Chim. France 5:813
117. Nakamura H, Ohizumi Y, Kobayashi J, Hirata Y (1984) Tetrahedron Lett. 25:2475
118. Walker RP, Faulkner DJ, Van Engen D, Clardy J (1981) J. Am. Chem. Soc. 103:6772
119. Chevolot L, Padua S, Ravi BN, Blyth PC, Scheuer PJ (1977) Heterocycles 7:891
120. Sharma GM, Burkholder PR (1971) J. Chem. Soc., Chem. Commun. 151
121. Sharma G, Magdoff-Fairchild B (1977) J. Org. Chem. 42:4118
122. Fedoreyev SA, Utkina NK, Ilyin SG, Reshetnyak MV, Maximov OB (1986) Tetrahedron Lett. 27:3177
123. Sharma GM, Buyer JS, Pomerantz MW (1980) J. Chem. Soc., Chem. Commun. 435
124. Kitagawa I, Kobayashi M, Kitanaka K, Kido M, Kyogoku Y (1983) Chem. Pharm. Bull. 31:2321
125. Schmitz FJ, Gunasekera SP, Lakshmi V, Tillekeratne LMV (1985) J. Nat. Prod. 48:47
126. Utkina NK, Fedoreev SA, Maksimov OB (1984) Khim. Prir. Soedin. 535; Chem. Abs. 102:146334h
127. Ireland C, Scheuer PJ (1980) J. Am. Chem. Soc. 102:5688
128. Ireland CM, Durso Jr. AR, Newman RA, Hacker MP (1982) J. Org. Chem. 47:1807
129. Hamamoto Y, Budd M, Nakagawa M, Nakanishi T, Mizukawa K (1983) J. Chem. Soc., Chem. Commun. 323
130. Wasylyk JM, Biskupiak JE, Costello CE, Ireland CM (1983) J. Org. Chem. 48:4445
131. Sesin DF, Gaskell SJ, Ireland CM (1986) Bull. Soc. Chim. Belg. 95:853
132. Biskupiak JE, Ireland CM (1983) J. Org. Chem. 48:2302
133. Hamada Y, Kato S, Shioiri T (1985) Tetrahedron Lett. 26:3223
134. Schmidt U, Utz R, Gleich P (1985) Tetrahedron Lett. 26:4367
135. Schmidt U, Griesser H (1986) Tetrahedron Lett. 27:163
136. Schmidt U, Weller D (1986) Tetrahedron Lett. 27:3495
137. Ireland CM unpublished results

138. Shioiri T, Hamada Y, Kato S, Shibata M, Kondo Y, Nakagawa H, Kohda K (1987) Biochemical Pharmacology 36:4181
139. Rinehart Jr. KL, Gloer JB, Hughes Jr. RG, Renis HE, McGovren JP, Swynenberg EB, Stgringfellow DA, Kuentzel Sl, Li LH (1981) Science 212:933
140. Rinehart Jr. Kl, Gloer JB, Cook Jr. JC, Mizsak SA, Scahill TA (1981) J. Am. Chem. Soc. 103:1857
141. Rinehart Jr. KL (1987) 28th Annual Meeting of the American Society of Pharmacognosy, Kingston, Rhode Island, July 1987.
142. Canonico PG, Pannier WL, Huggins JW, Rinehart Jr. KL (1982) Antimicrob. Agents Chemother. 22:696
143. Jiang TL, Liu RH, Salmon SE (1983) Cancer Chemother. Pharmacol 11:1
144. Montgomery DW, Zukoski CF (1985) Transplant 40:49
145. Bruening RC, Oltz EM, Furukawa J, Nakanishi K, Kustin K (1985) J. Am. Chem. Soc. 107:5298
146. Bruening RC, Oltz EM, Furukawa J, Nakanishi K, Kustin K (1986) J. Nat. Prod. 49:193
147. Kem W. personal communication
148. Schaller HC (1973) J. Embryol. Exp. Morphol. 29:27
149. Schaller HC, Bodenmuller H (1981) Proc. Natl. Acad. Sci. USA 78:7000
150. Grimmelikhijzen CJP (1984) Pepetides in the nervous system of coelenterates. In: Falkner S, Hakanson R, Sundler F (eds) Evolution and Tumor Pathology of the Neuroendocrine System, Elsevier, Amsterdam, chap 3
151. Grimmelikhuijken CJP, Graff D (1986) Proc. Natl. Acad. Sci. USA 83:9817
152. Wunderer G, Eulitz M (1978) Eur. J. Biochem. 89:11
153. Wunderer G, Machleidt W, Wachter E (1976) Hoppe-Seyler's Z. Physiol. Chem. 357:239
154. Scheffler JJ, Tsugita A, Linden G, Schweitz H, Luzdunski M (1982) Biochem. Biophys. Res. Commun. 197:272
155. Tanaka M, Haniu M, Yasunobu KT, Norton TR (1977) Biochem. 16:204
156. Reimer NS, Yasunobu CL, Yasunobu KT, Norton TR (1985) J. Biol. Chem. 260:8690
157. Kem WR, Dunn B, Parten B, Pennington M, Price D (1986) Fed. Proc. 45:1795
158. Zykova TA, Vinokurov LM, Koslovskaya EP, Elyakov GB (1985) Bioorg Khim 11:302; Chem. Abs. 102:224821f
159. Schweitz H, Bidard JN, Frelin C, Pauron D, Vijverberg HPM, Mahasneh DM, Lazdunski M, Vilbois F, Tsugita A (1985) Biochem. 24:3554
160. Wemmer DE, Kumar NV, Metrione RM, Lazdunski M, Drobny G, Kallenbach NR (1986) Biochem. 25:6842
161. Shibata S (1984) Seitai no Kagaku 35:417; Chem. Abs. 102:198986s
162. Martinez G, Kopeyan C, Schweitaz H, Lazdunski M (1977) FEBS Lett 84:247
163. Nishida S, Fujita S, Warashina A, Satake M, Tamiya N (1985) Eur. J. Biochem. 150:171
164. Narahashi T, Moore J, Shapiro BI (1969) Science 163:680
165. Catterall WA, Beress L (1978) J. Biol. Chem. 253:393
166. Vincent VP, Balerna M, Barhanin J, Fosset M, Lazdunski M (1980) Proc. Natl. Acad. Sci. USA 77:1646
167. Scriabne A, Van Arman CG, Morgan G, Morris AA, Bennett CD, Bohidar NR (1979) J. Cardiov. Pharmacol. 1:571
168. Catterall WA, Coppersmith J (1981) Mol. Pharmacol. 20:533
169. Fujita S, Warashina A, Sataka M (1983) Comp. Biochem. Physiol. 76C:25
170. Ishizaki H, McKay RH, Norton TR, Yasunobu KT, Lee J, Tu AT (1979) J. Biol. Chem. 254:9651

171. Nabiullin AA, Odinokov SE, Kozlovskaya EP, Elyakov GB (1982) FEBS Lett. 141:124
172. Gooley PR, Norton RS (1986) Biochem. 25:2349
173. Kem WR (1987) Proc. Cal. Acad. Sci. 13:69–84.
174. Mahon AC, Lloyd PE, Weiss KR, Kupfermann I, Scheller RH (1985) Proc. Natl. Acad. Sci. USA 82:3925
175. Rothman BS, Hawke DH, Brown RO, Lee TD, Dehghan AA, Shively JE, Mayeri E (1986) J. Biol. Chem. 261:1616
176. Petitt GR, Kamano Y, Brown P, Gust D, Inoue M, Gerald CL (1982) J. Am. Chem. Soc. 104:905
177. Hamada Y, Kohda K, Shioiri T (1984) Tetrahedron Lett. 25:5303
178. Schmidt U (1986) Pure Appl. Chem. 58:295
179. Pettit GR, Holzapfel CW (1986) J. Org. Chem. 51:4580
180. Pettit GR, Kamano Y, Holzapfel W, van Zyl WJ, Tuinman AA, Herald CL, Baczynskyj L, Schmidt JM (1987) J. Amer. Chem. Soc. 109:7581
181. Pettit GR, Kamano Y, Herald CL, Tuinman AA, Boettner FE, Kizu H, Schmidt JM, Baczynskyj L, Tomer KB, Bontems RJ (1987) J. Amer. Chem. Soc. 109:6883
182. Kamano T, Kizu J, Pettit GR, Herald CL, Tuinman AA, Bontems RL (1987) Tennen Yuki Kagobutsu Koen Yoshishu 29:295: Chem. Abs. 108:89927b
183. Sodano G, Spinella A (1986) Tetrahedron Lett. 27:2505
184. Sangster AW, Thomas SE, Tingling NL (1975) Tetrahedron 31:1135
185. Leung M, Stefano GB (1983) Life Sci 77, suppl. I
186. Greenberg MJ, Price DA (1979) Amer. Zool. 19:163
187. Price DA (1986) Amer. Zool. 26:1007
188. Greenberg MJ, Painter SD, Doble KE, Nagle GT, Price DA, Lehman HK (1983) Fed. Proc. 42:82
189. Cottrell GA, Davies NW (1987) J. Physiol. 382:51
190. Greenberg MJ (1987) Proc. Cal. Acad. Sci. in press
191. Price DA, Davies NW, Doble KE, Greenberg MJ (1987) Zool. Sci. 4:395
192. Blumenthal KM, Kem WL (1980) J. Biol. Chem. 255:8266
193. Blumenthal KM (1980) J. Biol. Chem. 255:8273
194. Blumenthal KM (1982) Biochem. 21:4229
195. Blumenthal KM, Keim PS, Heinrikson RL, Kem WR (1981) J. Biol. Chem. 256:9063
196. Blumenthal KM, Kem WR (1980) Arch. Biochem. Biophys. 203:816
197. Blumenthal KM (1980) Arch. Biochem. Biophys. 203:822
198. Hashimoto Y (1979) In: Marine toxins and other bioactive marine metabolites, Japan Scientific Societies Press, Tokyo, p 330
199. Zahuranec BJ (1983) Shark repellants from the sea; new perspectives. American Association for the Advancement of Science, Washington DC
200. Tachibana K, Sakaitani M, Nakanishi K (1984) Science 226:703
201. Tachibana K, Sakaitani M, Nakanishi K (1985) Tetrahedron 41:1027
202. Primor N, Parness J, Zlotkin E (1978) In: Rosenberg P (ed) Toxins: animal, plant and microbial, Pergamon, Oxford, p 539
203. Thompson SA, Tachibana K, Nakanishi K, Kubota I (1986) Science 233:341
204. Schroder E, Lubke K, Lehmann M, Beetz I (1971) Experimentia 27:764
205 Liguori YR, Ruggieri GD, Baslow MH, Stempien MF, Nigrelli RF (1963) Amer. Zool. 3:546
206. Maretzki A, Del Castillo J (1967) Toxicon 4:245
207. Randall JE, Arda K, Hibiya T, Mitsuura N, Kamiya H, Hasimoto Y (1971) Publs. Seto Mar. Biol. Lab. 19:157
208. Hashimoto Y, Oshima Y (1972) Toxicon 10:279

209. Hashimoto Y, Shiomi K, Aida K (1974) Toxicon 12:523
210. Hori K, Fusetani N, Hashimkoto K, Aida K, Randall JE (1979) Toxicon 17:418
211. Oshima Y. Hashimoto Y (1976) In: Oshima Y, Hashimoto Y (eds) Animal, plant and microbial toxins. Plenum, New York, Vol 1, p 303

Sperm-Activating Peptides from Sea Urchin Egg Jelly

Norio Suzuki

Contents

Abstract

Speract (Gly-Phe-Asp-Leu-Asn-Gly-Gly-Gly-Val-Gly) and ten of its derivatives obtained from the egg jelly of sea urchins, Order Echinoida, stimulate the respiration and motility of spermatozoa of sea urchins belonging to the same order. Resact (Cys-Val-Thr-Gly-Ala-Pro-Gly-Cys-Val-Gly-Gly-Gly-Arg-LeuNH_2) and mosact (Asp-Ser-Asp-Ser-Ala-Gln-Asn-Leu-Ile-Gly) and two of its derivatives obtained from *Arbacia punctulata* eggs, Order Arbacioida, and *Clypeaster japonicus* eggs, Order Clypeasteroida, also stimulate the metabolism and motility of respective sea urchin spermatozoa. Two other peptides (Gly-(X)-Pro-(X)-Gly-Gly-Ala-Val- and Gly-Cys-Pro-Trp-Gly-Gly-Ala-Val-Cys) were isolated from the egg jelly of the sea urchin *Diadema setosum*, Order Diadematoida. These are specific for *Diadema setosum* spermatozoa. Early biochemical responses of the spermatozoa to the egg jelly peptides include at a net hydrogen ion efflux and elevations of cyclic AMP and cyclic GMP concentrations. In additoin, a major sperm plasma protein is modified in response to the peptides such that its apparent molecular weight is lowered with exception of the response of

---

Noto Marine Laboratory, Kanazawa University, Ogi, Uchiura, Ishikawa 927-05, Japan

Bioorganic Marine Chemistry, Volume 3

Glyptocidaris crenularis spermatozoa to alloresact (Lys-Leu-Cys-Pro-Gly-Gly-Asn-Cys-Val). The plasma membrane protein appears to be the enzyme guanylate cyclase. Resact acts as a chemoattractant for *Arbacia punctulata* and speract promotes the acrosome reaction of *Hemicentrotus pulcherrimus* spermatozoa as a cofactor.

1 Introduction

It has been known for over 60 years that soluble factors associated with the eggs of certain species of sea urchins enhance the respiration and motility of sea urchin spermatozoa [1]. The factors have been reported to be diffusible in dialysis, heat-stable, alcohol-soluble, and non-volatile [2]. In 1976, Ohtake demonstrated that the respiration of sea urchin spermatozoa could be reproducibly stimulated by partially purified egg jelly factors obtained from the sea urchin *Pseudocentrotus depressus* if the extracellular pH was maintained at acidic values [3–4]. Thereafter, Kopf *et al.* found that the egg jelly of the sea urchins *Strongylocentrotus purpuratus* and *Lytechinus pictus* contains a factor which elevates the respiratory rate and cAMP and cGMP concentrations of homologous spermatozoa [5]. The molecular weight of the factor obtained from *Strongylocentrotus purpuratus* was estimated to be 1800 daltons by gel filtration [6]. Hansbrough and Garbers named the factor speract which stands for sperm activating peptide [6]. In 1981, we purified the factor from the egg jelly of the sea urchin *Hemicentrotus pulcherrimus* and demonstrated that the factor is a decapeptide whose structure is Gly-Phe-Asp-Leu-Asn-Gly-Gly-Gly-Val-Gly [7]. This contribution describes the structure and function of sperm-activating peptides.

2 Purification and Structure

The sea urchin egg is surrounded by a transparent extracellular envelope known as the jelly layer, through which spermatozoa must pass before reaching the plasma membrane of the egg during fertilization. The jelly layer is mainly composed of a sialopolysaccharide-protein complex, which could be separated into a sialoglycoprotein and a fucose sulfate polysaccharide [8–12]. Considering the high content of carboxyl residues in the bound sialic acids of the sialoglycoprotein and acidic groups of ester-linked sulfate in the fucose sulfate polysaccharide, pH values within the jelly layer seem to be lower than the pH of normal sea water (pH 8.2) [13]. In fact, pH values of centrifugally isolated jelly layer were 0.4 to 0.5 units lower than that of fresh normal sea water (pH 8.2). Recently, Ciccarelli *et al.* demonstrated that pH values of starfish jelly layer were between 6.6 and 7.2 [14]. At low pH values, the respiration and motility of sea urchin spermatozoa are markedly reduced (Fig. 1) and their movements in the jelly layer, therefore, are slower than those that prevail in sea water. Thus, it is predicted that the egg jelly is provided with a system of controlling motility and respiration, because the spermatozoa should maintain their activities at the initial high level attained in sea water during their passage throgh the acidic environment of the jelly layer to reach the egg surface. Because of

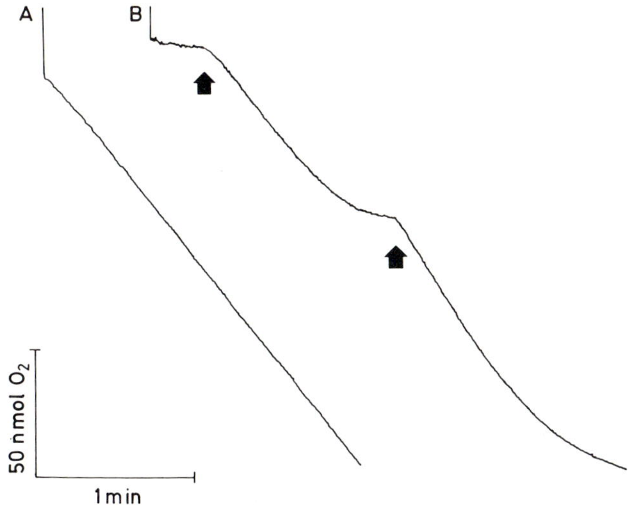

Fig. 1. Respiration of *Hemicentrotus pulcherrimus* spermatozoa in artificial sea water (A) pH 8.2, (B) pH 6.8. At the arrows, synthetic egg jelly peptide (speract) was added in the final concentration of 3.3×10^{-9} M

high viscosity, it is quite difficult to handle an intact jelly layer by common biochemical methods. To analyze the jelly layer, eggs are treated with sea water adjusted to pH 5.0, which dissolves the jelly layer. The centrifuged supernatant 'soluble egg jelly' is then mixed with a 2-fold volume of ethanol to extract peptides and then centrifuged. The resultant supernatant is concentrated under reduced pressure at 50° C and lipids are removed by chloroform extraction. The water layer is then lyophilized and the residue, after being dissolved in distilled water, is used for peptide purification.

2.1 Speract and its Derivatives

When we first isolated the sperm-activating peptide from the egg jelly of the sea urchin *Hemicentrotus pulcherrimus*, we used sequential chromatographies on Sephadex G-25, DEAE-Sephadex A-25, Sephadex G-15 and Avicel SF thin layer plate [7]. As shown in Fig. 2 and Table 1, we obtained two peptides (H-1: Gly-Phe-Asp-Leu-*Thr*-Gly-Gly-Gly-Val-Gly; H-2: Gly-Phe-Asp-Leu-*Asn*-Gly-Gly-Gly-Val-Gly). These peptides were quite potent in stimulating *Hemicentrotus pulcherrimus* sperm respiration that had been lowered under slightly acidic conditions (pH 6.8), but the peptide-stimulated respiration rates did not exceed those attained in normal sea water even when a high concentration of peptide was employed. At that time we used about 20 liters of soluble egg jelly prepared from 5,000 female sea urchins and obtained a total of 15 mg of pure peptides (7 mg H-1 and 8 mg H-2). Since the biological activity of the peptide H-2 was similar to that of substance obtained from the egg jelly of *Strongylocentrotus purpuratus*, the peptide was named speract and the

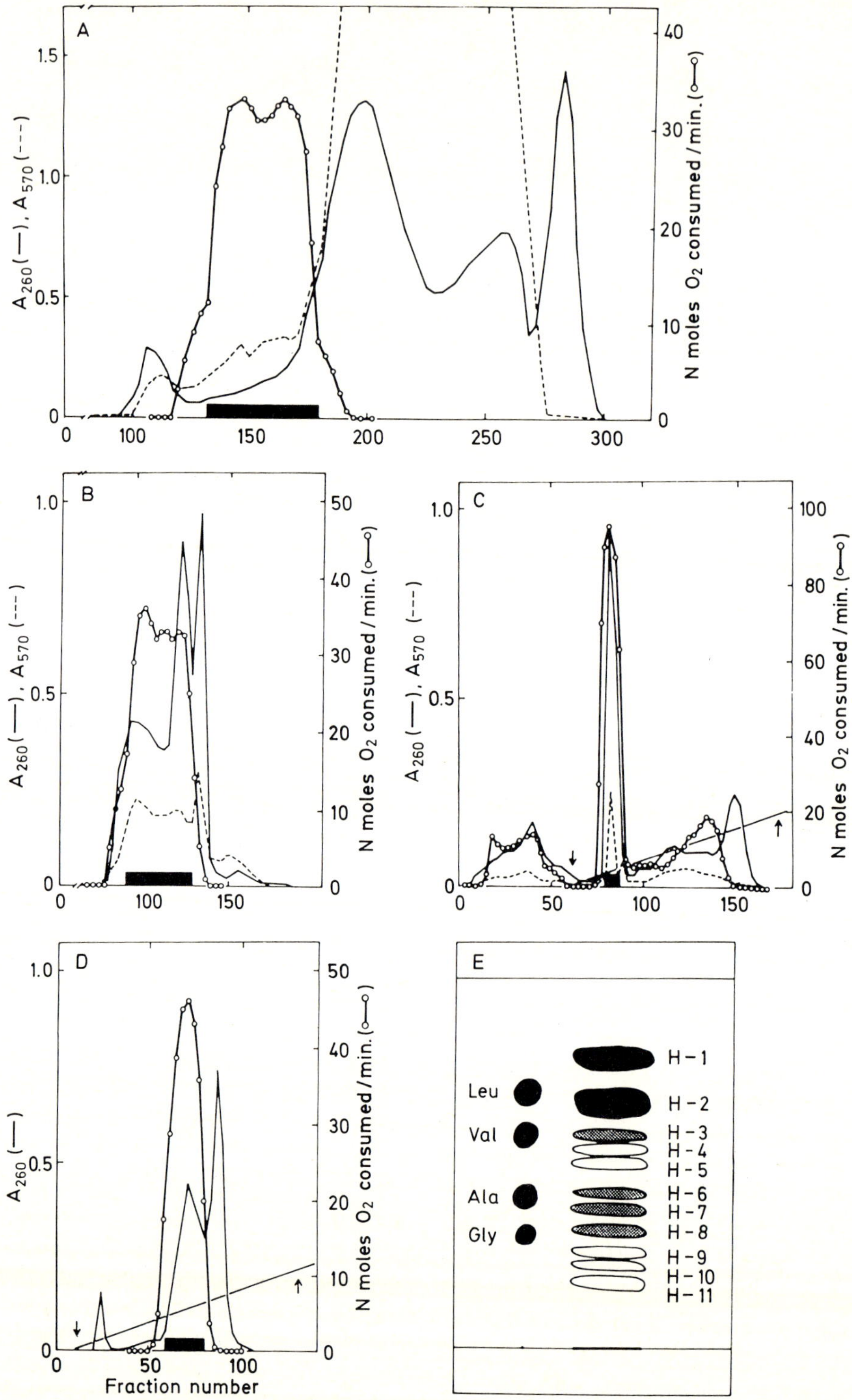

Fig. 2. Continued

F

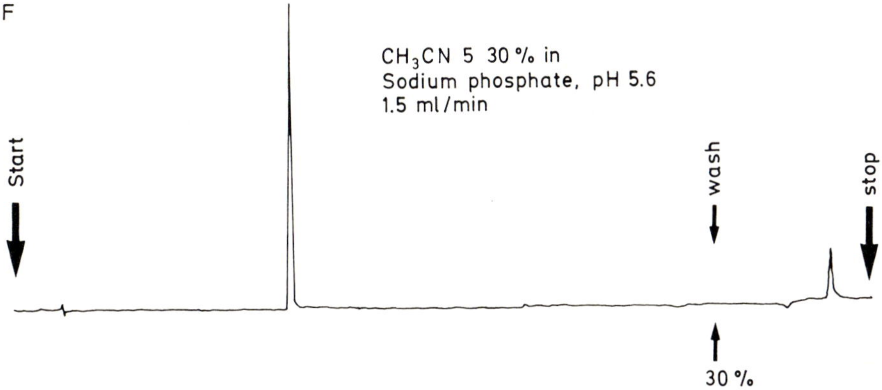

Fig. 2. Elution profiles of sperm activating peptides from the egg jelly of the sea urchin *Hemicentrotus pulcherrimus* by sequential chromatographies on (A) Sephadex G-25, (B) Sephadex G-15, (C) DEAE-Sephadex A-25, (D) DEAE-Ssephadex A-25 and (E) Avicel SF thin layer plate. (F) a HPLC profile of synthetic H-2 (speract) on a reverse-phase column

Table 1. The amino acid composition[a], sequence, and molecular weight of the peptides from the egg jelly of the sea urchin *Hemicentrotus pulcherrimus*

Amino acid	H-1	H-2
Asp	0.99 (1)	1.85 (2)
Thr	0.88 (1)	–
Gly	4.96 (5)	4.55 (5)
Val	1.17 (1)	1.01 (1)
Leu	1.00 (1)	1.00 (1)
Phe	1.01 (1)	0.95 (1)
Total residue	(10)	(10)

Sequence and molecular weight of peptides
H-1 Gly-Phe-Asp-Leu-Thr-Gly-Gly-Gly-Val-Gly (Mr 879)
H-2 Gly-Phe-Asp-Leu-Asn-Gly-Gly-Gly-Val-Gly (Mr 892)

[a] The compositions are shown as normalized values with leucine to have one residue and numbers in parentheses refer to the most probable number of residues in peptides

peptide H-1 was named as a speract derivative ([Thr^5]-speract) [15]. This procedure, however, was time-consuming and required a large amount of soluble egg jelly to obtain pure peptides. We then modified the purification procedure and now we use sequential HPLC on a reverse-phase column to purify the peptides [16]. In general, separations are carried out using a combination of the following programs: Program I: Flow rate is 9.9 ml/min, the column (Shim-pack C-8 Prep, particle size 5 μm, 20 × 250 mm) is equilibrated with 5% acetonitrile (ACN) in 0.1% trifluoroacetic acid (TFA) in water and eluted for 15 min with equilibration solvent, followed by elution with 60% ACN on 0.1% TFA in water for the next 15 min. Program II: Flow rate is 1.0 ml/min, the column (Unisil C-8, particle size 5 μm, 4.6 × 250 mm) is equilibrated with 10% ACN in 0.1% TFA and eluted for 10 min with equilibration solvent, followed by a linear gradient, of ACN from 10% to 50% in 0.1% TFA over a 50 minute time period. Program III: Flow rate is 1.0 ml/min, the same column used in Program II is equilibrated with 5% ACN in 5 mM sodium phosphate (pH 5.7) and eluted for 20 min with the equilibration solvent, followed by a linear gradient of ACN from 5% to 30% in 5 mM sodiumn phosphate (pH 5.7) over a 40 minute time period. By using the modified procedure, we obtained three new speract derivatives ([$Ser^{3,10}$]-speract, [Ser^1, Ala^3, Gly^5]-speract and [$Ser^{3,5,7}$, Asp^{10}]-speract) in addition to speract and [Thr^5]-speract from the egg jelly of 150 female *Hemicentrotus pulcherrimus* sea urchins (see Fig. 3 and Table 2) [17]. With essentially the same purification method as with *Hemicentrotus pulcherrimus*, we purified eleven sperm activating peptides from four other species (*Lytechinus pictus, Pseudocentrotus depressus, Strongylocentrotus purpuratus* and *Anthocidaris crassispina*) of the order Echinoida to which *Hemicentrotus pulcherrimus* also belongs [(17-19)]. From the egg jelly of 100 female *Pseudocentrotus depressus*, 175 nmol of speract, 188 nmol of [Thr^5]-speract and

Table 2. The amino acid composition[a] and sequence of sperm-activating peptides obtained from the egg jelly of the sea urchin *Hemicentrotus pulcherrimus*

Amino acid	Peptide a	Peptide b	Peptide c	Peptide d	Peptide e
Asp	1.97 (2)	2.03 (2)	–	1.03 (1)	1.02 (1)
Thr	–	–	–	1.02 (1)	0.98 (1)
Ser	0.94 (1)	–	0.99 (1)	–	1.94 (2)
Gly	3.88 (4)	4.83 (5)	4.95 (5)	4.89 (5)	2.98 (3)
Ala	–	–	1.00 (1)	–	–
Val	1.00 (1)	1.00 (1)	1.00 (1)	1.00 (1)	1.00 (1)
Leu	1.01 (1)	1.05 (1)	1.05 (1)	1.06 (1)	1.03 (1)
Phe	1.02 (1)	1.03 (1)	1.03 (1)	1.02 (1)	1.01 (1)
Total residue	(10)	(10)	(10)	(10)	(10)

Sequence of peptides
Speract:
Gly-Phe-Asp-Leu-Asn-Gly-Gly-Gly-Val-Gly (peptide b)
[Thr^5]-Speract:
Gly-Phe-Asp-Leu-Thr-Gly-Gly-Gly-Val-Gly (peptide d)
[$Ser^{3,10}$]-Speract:
Gly-Phe-Ser-Leu-Asn-Gly-Gly-Gly-Val-Ser (peptide a)
[Ser^1, Ala^3, Gly^5]-Speract:
Ser-Phe-Ala-Leu-Gly-Gly-Gly-Gly-Val-Gly (peptide c)
[$Ser^{3,5,7}$, Asp^{10}]-Speract:
Gly-Phe-Ser-Leu-Ser-Gly-Ser-Gly-Val-Asp (peptide e)

[a] The compositions are shown as normalized values with valine to have one residue and numbers in parentheses refer to the most probable number of residues in peptides

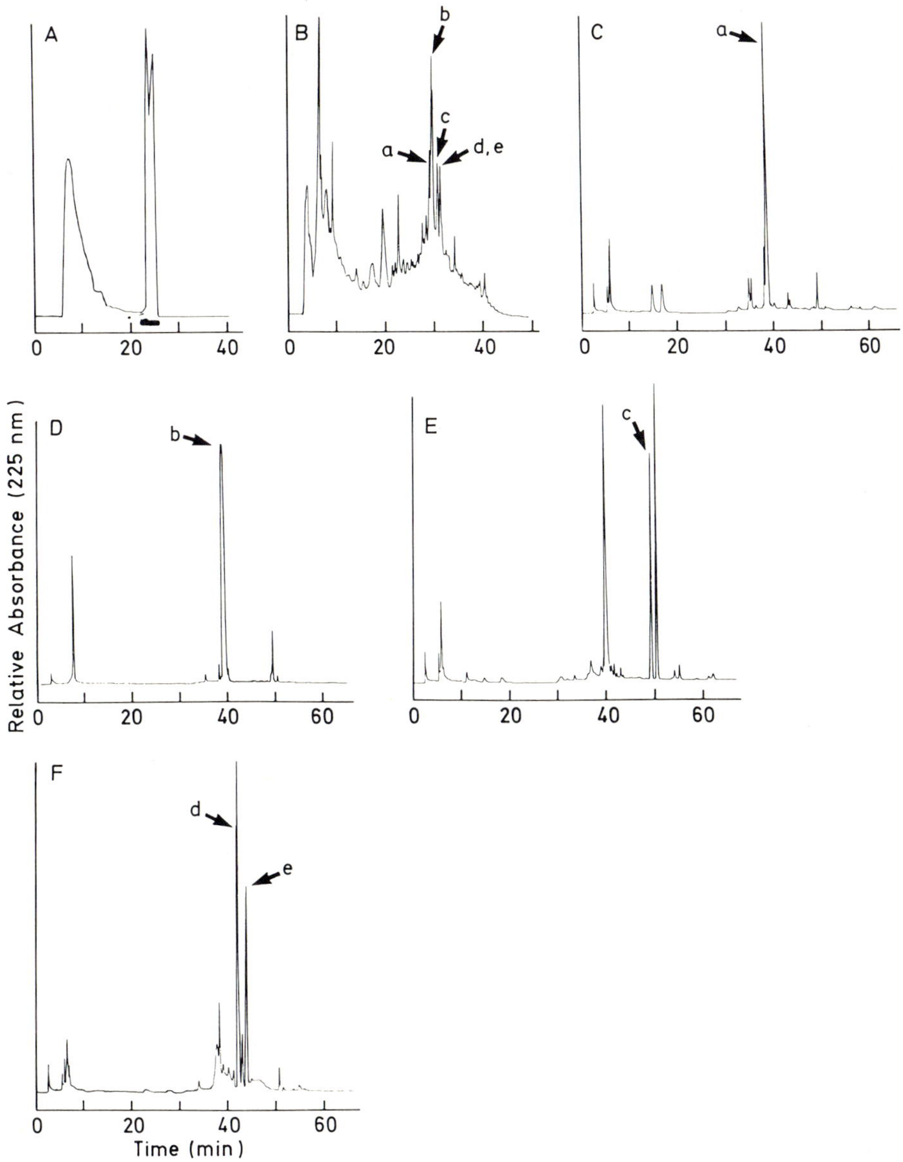

Fig. 3. HPLC profiles of *Hemicentrotus pulcherrimus* egg jelly peptides by Program I (A), Program II (B) and Program III (D–F)

186 nmol of [Ala^3, Gly^5]-speract were obtained [17]. Three peptides stimulated sperm respiration of *Pseudocentrotus depressus* at half of the maximum level in the range of 0.8 nM to 3 nM. Although the values were a little less than those for *Hemicentrotus pulcherrimus* spermatozoa, the peptides were still quite potent stimulators for *Pseudocentrotus depressus* sperm respiration.

In addition to speract, *Strongylocentrotus purpuratus* egg jelly contained two speract derivatives, [Ala^3, Gly^5]-speract and [Ser^3, Thr^5]-speract. Similarly, *Anthocidaris crassispina* contains [$Ser^{3,5,7}$]-speract as well as [Ser^5]-speract and [Thr^5]-speract [18]. *Lytechinus pictus* contained a speract derivative lacking the NH_2-terminal Gly and the derivative exhibited almost the same respiration stimulating activity toward homologous and *Hemicentrotus pulcherrimus* spermatozoa [19]. The derivative does not seem to orginate directly from the long peptide in the egg jelly, because the long peptide was recovered in greater than 95% yield from the incubation mixture of the peptide and crude mixtures of egg jelly [19].

These five species belonging to the order Echinoida have speract and/or speract derivatives of which the general structure can be written as X_1-Phe-X_2-Leu-X_3-Gly-X_4-Gly-Val-X_5 (X_1 is Gly or Ser; X_2 is Asp, Ala or Ser; X_3 is Asn, Thr, Ser, Ala or Gly; X_4 is Gly or Ser; X_5 is Gly, Ser, Gln or Asp) (Fig. 4). Speract and all of its derivatives activate *Hemicentrotus pulcherrimus* spermatozoa as well as homologous spermatozoa at one half of maximum at about 8 to 60 pM [17]. Although speract and its derivatives are effective in stimulating the respiration of *Hemicentrotus pulcherrimus* spermatozoa as well as sperm respiration of sea urchins in the order Echinoida, they do not stimulate sperm respiration of sea urchins in different taxonomic orders [20–21].

```
Order Echinoida
 Suborder Temnopleurina
   Family Toxopneustidae
             Lytechinus pictus
                 [Thr5,Gln10]-Speract: Gly-Phe-Asp-Leu-Thr-Gly-Gly-Gly-Val-Gln
        Des-Gly1-[Thr5,Gln10]-Speract:     Phe-Asp-Leu-Thr-Gly-Gly-Gly-Val-Gln
             Pseudocentrotus depressus
                              Speract: Gly-Phe-Asp-Leu-Asn-Gly-Gly-Gly-Val-Gly
                       [Thr5]-Speract: Gly-Phe-Asp-Leu-Thr-Gly-Gly-Gly-Val-Gly
                  [Ala3,Gly5]-Speract: Gly-Phe-Ala-Leu-Gly-Gly-Gly-Gly-Val-Gly
 Suborder Echinina
   Family Strongylocentrotidae
             Strongylocentrotus purpuratus
                              Speract: Gly-Phe-Asp-Leu-Asn-Gly-Gly-Gly-Val-Gly
                  [Ala3,Gly5]-Speract: Gly-Phe-Ala-Leu-Gly-Gly-Gly-Gly-Val-Gly
                  [Ser3,Thr5]-Speract: Gly-Phe-Ser-Leu-Thr-Gly-Gly-Gly-Val-Gly
             Hemicentrotus pulcherrimus
                              Speract: Gly-Phe-Asp-Leu-Asn-Gly-Gly-Gly-Val-Gly
                       [Thr5]-Speract: Gly-Phe-Asp-Leu-Thr-Gly-Gly-Gly-Val-Gly
                    [Ser3,10]-Speract: Gly-Phe-Ser-Leu-Asn-Gly-Gly-Gly-Val-Ser
             [Ser1,Ala3,Gly5]-Speract: Ser-Phe-Ala-Leu-Gly-Gly-Gly-Gly-Val-Gly
             [Ser3,5,7,Asp10]-Speract: Gly-Phe-Ser-Leu-Ser-Gly-Ser-Gly-Val-Asp
   Family Echinometridae
             Anthocidaris crassispina
                       [Thr5]-Speract: Gly-Phe-Asp-Leu-Thr-Gly-Gly-Gly-Val-Gly
                       [Ser5]-Speract: Gly-Phe-Asp-Leu-Ser-Gly-Gly-Gly-Val-Gly
                    [Ser3,5,7]-Speract: Gly-Phe-Ser-Leu-Ser-Gly-Ser-Gly-Val-Gly
```

Fig. 4. Structural similarity of sperm activating peptides of five species in the order Echinoida

2.2 Resact, Alloresact and its Derivatives

It has been reported that the egg jelly of the sea urchin *Arbacia punctulata*, which belongs to the order Arbacioida, lacks the agent capable of stimulating the sperm respiration of *Lytechinus variegatus* in the order Echinoida despite a similarity in the gross chemical composition of the egg jelly of both species [2, 12]. This suggests that, by means of sperm-activating peptides or factors, there is species specificity at the ordinal level in the respiratory stimulation of sea urchin spermatozoa. To confirm this, we initially determined that speract did not activate the respiration rates of *Arbacia punctulata* spermatozoa. We then determined that the egg jelly from *Arbacia punctulata* contained a substance(s) which stimulates the respiration of its own spermatozoa. This substance did not activate *Strongylocentrotus purpuratus* or *Lytechinus pictus* spermatozoa. Purification of the substance by HPLC demonstrated that it was a peptide [22]. The amino acid composition of the peptide after being treated with [^{14}C]iodoacetic acid under both reducing and nonreducing conditions was as follows: Thr(0.9 nmol), Pro(0.8 nmol), Gly(4.6 nmol), Ala(0.8 nmol), Val(1.8 nmol), Leu(1.0 nmol), CmCys(2.2 nmol) and Arg(1.0 nmol). The carboxymethylated peptide was subjected to sequence analysis and the following sequence was determined by Edman degradataion: CmCys-Val-Thr-Gly-Ala-Pro-Gly-CmCys-Val-Gly-Gly-Gly-Arg. No Leu was detected at the COOH-terminus. To determine the COOH-terminal amino acid, 2.8 nmol of the peptide were incubated with 10 μg of carboxypeptidase Y in 10 nM NH_4HCO_3 at 30°C for 4.5 h and the solution was then lyophilized. The residue was applied to the amino acid analyzer and Leu was detected as the released amino acid (50–100% yield). To confirm that Leu is the COOH-terminal amino acid, the peptide was synthesized with Leu as the COOH-terminal amino acid. When the synthesized peptide was compared to a native peptide by HPLC, however, the two peptides migrated differently by approximately 2 min on reverse-phase (octyl) columns. Thus, the leucine-amide form of the peptide was prepared as follows: amidation was performed by reacting the peptide with ammonium chloride (5.5 M NH_4Cl, pH 4.75) in the presence of 1-ethyl-3-dimethylaminopropylcarbodiimide. The peptide containing leucine-amide as the COOH-terminus migrated identically on reverse-phase columns and on thin layer chromatography with the native peptide [23]. Both the synthesized peptide with leucine-amide and native peptide elevated *Arbacia punctulata* sperm respiration at half of the maximum rate at 0.5 nM [23]. The peptide, obviously different from speract and its derivatives based on both amino acid sequence as well as biological activity, has been named resact which stands for respiration-activating peptide [22–23].

The sea urchin *Glyptocidaris crenularis* belongs to the order Arbacioida in which *Arbacia punctulata* is also included. From the egg jelly from 50 female *Glyptocidaris crenularis*, six sperm-activating peptides were purified by repeated HPLC and sequenced by Edman degradation [24]. At the sequence analysis, PTH-cysteine was not detected. But amino acid analysis data indicated that all six peptides have two residues of cysteine. To confirm the validity of the sequenced structure, we chemically synthesized the peptides which have cysteine at the corresponding positions. The synthesized peptides were eluted at the same retention time as the respective native peptide and also elevated *Glyptocidaris crenularis* sperm respiration at the same concentration as the native peptides. The six native peptides and synthesized peptides stimulated *Glyptocidaris crenularis* sperm respiration at one half of the maximum rate at concentration between 0.5 nM and 1 nM [24]. Resact, structurally unrelated to the peptides obtained from *Arbacia punctulata*, also stimulated *Glyptocidaris crenularis* sperm respiration although half-maximal stimulation of *Glyptocidaris crenularis* sperm respiration required 500 nM of resact. The peptide, Lys-Leu-Cys-Pro-Gly-Gly-Asn-Cys-Val, was obtained from the egg jelly in the largest amount (240 nmol). Therefore, the peptide was named alloresact which means structurally different from resact and biologically similar to resact. The other five peptides were named as alloresact derivatives [24].

2.3 Mosact and its Derivatives

The egg jelly of the sea urchin *Clypeaster japonicus* in the order Clypeasteroida contains three structurally similar sperm-activating peptides [16]. The peptides stimulated *Clypeaster japonicus* sperm respiration at one half of the maximum rate at about 0.5 nM. The peptides, speract and resact as well as alloresact, do not cross-react. Therefore, the peptide Asp-Ser-Asp-Ser-Ala-Gln-Asn-Leu-Ile-Gly, from *Clypeaster japonicus* was named mosact which stands for motility-activating peptide and the other two peptides were named as mosact derivatives [16]. When the peptide Des-Gln^6, Asn^7-[Phe^6]-mosact was subjected to sequence analysis, the sixth cycle resulted in release of no identifiable PTH-amino acid and PTH-Phe was not detected at any cycle. The sixth residue of the peptide was assigned to Phe since the amino acid analysis data clearly showed that peptide contained one Phe redsidue and a chemically synthesized peptide which has Phe at the sixth residue stimulated *Clypeaster japonicus* sperm respiration rates at the same concentration as the corresponding native peptide [16].

2.4 Other Peptides

As we have described so far, we purified twenty-six sperm-activating peptides from eight species distributed over three taxonomic orders (see Table 3). Considering the structure and biological specificity, the peptides could be classified in three groups, that is, speract, resact (including alloresact), and mosact. Therefore, a theory can be presented to suggest that peptides isolated from the eggs of on taxonomic order interact with sperm cells of other species within the same order, but not within other orders. The theory may extend to a theory that the apparent coevolution of peptide and receptor is due to the necessity of the resultant physiological responses. For this, we shall continue further investigations about peptides from the egg jelly of sea urchins in different orders.

Sea urchins in the order Diadematoida are considered to be of a more primitive type. From the egg jelly of 40 female *Diadema setosum*, Order Diadematoida, we purified two peptides by HPLC on reverse-phase columns [25]. The peptides stimulated *Diadema setosum* sperm respiration at one half of the maximum rate at about 100 pM. Speract, resact (also alloresact), and mosact did not stimulate the

Table 3. Sperm-activating peptides from the egg jelly of various species of sea urchins

- Subclass Regularia
  - Order Diadematodia
    - Suborder Diademina
      - Family Diadematidae
        - *Diadema Setosum*
        - Gly-(X)-Pro-(X)-Gly-Gly-Ala-Val-
        - Gly-Cys-Pro-Trp-Gly-Gly-Ala-Val-Cys[2]

- Order arbacioida
  - Suborder Phymosomatina
    - Family Phymosomatidae
      - *Glyptocidaris Crenularis*
      - Ser-Ala-Alloresact:
      - Ser-Ala-Lys-Leu-Cys-Pro-Gly-Gly-Asn-Cys-Val
      - Alloresact:
      - Lys-Leu-Cys-Pro-Gly-Gly-Asn-Cys-Val
      - Des-Lys^{1}-Alloresact:
      - Leu-Cys-Pro-Gly-Gly-Asn-Cys-Val
      - Ser,Phe-[Gln^{7}]-Alloresact:
      - Ser-Phe-Lys-Leu-Cys-Pro-Gly-Gly-Gln-Cys-Val
      - [Gln^{7}]-Alloresact:
      - Lys-Leu-Cys-Pro-Gly-Gly-Gln-Cys-Val
      - Des-Lys^{1}-[Gln^{7}]-Alloresact:
      - Leu-Cys-Pro-Gly-Gly-Gln-Cys-Val
  - Suborder Arbacina
    - Family Arbaciidae
      - *Arbacia Punctulata*
      - Resact:
      - Cys-Val-Thr-Gly-Ala-Pro-Gly-Cys-Val-Gly-Gly-Gly-Arg-Leu-NH_2

- Order echinoida
  - Suborder Temnopleurina
    - Family Toxopneustidae
      - *Lytechinus Pictus*
      - Des-Gly^{1}-[Thr^{5},Gln^{10}]-Speract:
      - Phe-Asp-Leu-Thr-Gly-Gly-Gly-Val-Gln
      - [Thr^{5},Gln^{10}]-Speract:
      - Gly-Phe-Asp-Leu-Thr-Gly-Gly-Gly-Val-Gln
      - *Pseudocentrotus Depressus*
      - Speract:
      - Gly-Phe-Asp-Leu-Asn-Gly-Gly-Gly-Val-Gly
      - [Thr^{5}]-Speract:
      - Gly-Phe-Asp-Leu-Thr-Gly-Gly-Gly-Val-Gly
      - [Ala^{3},Gly^{5}]-Speract:
      - Gly-Phe-Ala-Leu-Gly-Gly-Gly-Gly-Val-Gly

Table 3. (continued)

Suborder Echinina

Family Strongylocentrotidae

Strongylocentrotus purpuratus

Speract:

Gly-Phe-Asp-Leu-Asn-Gly-Gly-Gly-Val-Gly

[Ala^{3},Gly^{5}]-Speract:

Gly-Phe-Ala-Leu-Gly-Gly-Gly-Gly-Val-Gly

[Ser^{3},Thr^{5}]-Speract:

Gly-Phe-Ser-Leu-Thr-Gly-Gly-Gly-Val-Gly

Hemicentrotus pulcherrimus

Speract:

Gly-Phe-Asp-Leu-Asn-Gly-Gly-Gly-Val-Gly

[Thr^{5}]-Speract:

Gly-Phe-Asp-Leu-Thr-Gly-Gly-Gly-Val-Gly

[Ser^{3,10}]-Speract:

Gly-Phe-Ser-Leu-Asn-Gly-Gly-Gly-Val-Ser

[Ser^{1},Ala^{3},Gly^{5}]-Speract:

Ser-Phe-Ala-Leu-Gly-Gly-Gly-Gly-Val-Gly

[Ser^{3,5,7},Asp^{10}]-Speract:

Gly-Phe-Ser-Leu-Ser-Gly-Ser-Gly-Val-Asp

Family Echinometridae

Anthocidaris crassispina

[Thr^{5}]-Speract:

Gly-Phe-Asp-Leu-Thr-Gly-Gly-Gly-Val-Gly

[Ser^{5}]-Speract:

Gly-Phe-Asp-Leu-Ser-Gly-Gly-Gly-Val-Gly

[Ser^{3,5,7}]-Speract:

Gly-Phe-Ser-Leu-Ser-Gly-Ser-Gly-Val-Gly

Subclass Irregularia

Order Clypeasteroida

Suborder Clypeasterina

Family Clypeasteroidae

Clypeaster japonicus

Mosact:

Asp-Ser-Asp-Ser-Ala-Gln-Asn-Leu-Ile-Gly

Des-Gln^{6},Asn^{7}-[His^{6}]-Mosact:

Asp-Ser-Asp-Ser-Ala-His-Leu-Ile-Gly

Des-Gln^{6},Asn^{7}-[Phe^{6}]-Mosact:

Asp-Ser-Asp-Ser-Ala-Phe-Leu-Ile-Gly

[a] This structure was recently determined by a mass spectrometric method [25].

respiration of *Diadema setosum* spermatozoa. Sequence analysis data indicated that the sequence of peptide-1 is Gly-X-Pro-X-Gly-Gly-Ala-Val and that of petide-2 is Gly-X-Pro-Trp-Gly-Gly-Ala-Val. Amino acid composition of peptide-1 is as follows: Gly(3), Pro(1), Ala(1), Cys(2), Val(1) and that of peptide-2 is Gly(3), Pro(1), Ala(1), Cys(2), and Val(1). Spectroscopic measurement of the peptide-2 indicated that the peptide has one residue of tryptophan. Peptide-1 contains two residues of cysteine and peptide-2 has one residue of cysteine. Therefore, it is presumed that the unidentified amino acids are cysteine. Since the peptides are structurally not related to already purified sperm-activating peptides and biologically quite specific for *Diadema setosum* spermatozoa, a specific name should be given to the peptides, in the same way in which we have named speract, resact, and mosact. However, it is quite difficult to find a good name for these new peptides using this system. It is possible that we obtain very different peptides from the egg jelly of sea urchins in diffrent orders specific for given spermatozoa. These peptides as well as peptides with established structures should have essentially the same biological activities toward given spermatozoa as I will describe later. Therefore, we need a generalized name or an abbreviated name for these peptides. I propose sperm-activating peptide as the general name and SAP as the abbreviation. Thus, for example, speract and [Ser^5]-speract will be called sperm-activating peptide Type-I, which will be abbreviated as SAP-I, and [Ser^5]-sperm-activating peptide Type-I which is also abbreviated as [Ser^5]-SAP-I. In this nomenclature, resact becomes SAP-II and mosact SAP-III. When a new specific peptide such as the peptides from *Diadema setosum* egg jelly is found, it will be SAP-IV.

3 Chemical Synthesis and Chemical Nature of Sperm-Activating Peptides

Chemical syntheses of sperm-activating peptides and their analogs were carried out by the liquid phase method using CH_2Cl_2 as solvent with DCC or ClCOOIBu as a coupling reagent [26]. The protecting group Boc- was released by 2N HCl in dioxane/AcOEt (1:2), and the peptide ester-HCl was neutralized with *N*-methylmorpholine. The final coupling products with protecting groups, Cbz- and -OBzl were dissolved in methanol and deprotected by catalytic hydrogenation with palladium black as catalyst with bubbling of H_2 gas or HF treatment. The lyophilized free peptide was purified by repeated HPLC on a reverse-phase column.

Speract and its analogs with substitution and deletion synthesized are listed in Tables 4 to 6 [26]. Deletion of COOH-terminal Gly reduced the activity to about 1/3000, and the peptide identical to the chymotryptic fragment of [Ser^5]-speract has its activity rescued to 10^{-5}. However, the NH_2-terminal Gly could be removed without significant loss of activity. This implies that the minimal chain length of speract essential for the activity is nine residues. In contrast to the remarkable effects of deletion, modification at NH_2- or COOH-terminus has a smaller effect on the activity. Substitution by Pro usually reduced the activity. This may be due to the conformational change unfavorable for binding of the peptide to the receptor, while COOH-terminal Pro surprisingly increased the activity by 500-fold. Substitutoins by D-isomers also reduced the activity markedly except for D-Ser^5, which made the peptide 3.8 times more active than [Ser^5]-speract.

CD spectra of speract in H_2O and in 0.5 M NaCl have a maximum at 219 nm and a minimum at 229 nm (see Fig. 5). These spectra suggest that speract in

Table 4. Respiratory stimulation activity of synthetic [Ser^5]-speract analogs with different chain length[a]

Sequence	Relative activity
Gly-Phe-Asp-Leu-Ser-Gly-Gly-Gly-Val-Gly	1.0
Gly-Gly-Phe-Asp-Leu-Ser-Gly-Gly-Gly-Val-Gly	2.6
FTC-Gly-Phe-Asp-Leu-Ser-Gly-Gly-Gly-Val-Gly	1.9
Gly-Phe-Asp-Leu-Thr-Gly-Gly-Gly-Val-Gly	0.80
Ac-Gly-Phe-Asp-Leu-Ser-Gly-Gly-Gly-Val-Gly	0.25
Phe-Asp-Leu-Ser-Gly-Gly-Gly-Val-Gly	0.10
Gly-Phe-Asp-Leu-Ser-Gly-Gly-Gly-Val-Gly-OEt	4.8×10^{-2}
Gly-Phe-Asp-Leu-Thr-Gly-Gly-Gly-Val	3.5×10^{-4}
Gly-Asp-Leu-Ser-Gly-Gly-Gly-Val-Gly	1.8×10^{-4}
Asp-Leu-Ser-Gly-Gly-Gly-Val-Gly	2.9×10^{-5}
Asp-Leu-Ser-Gly-Gly-Gly-Val	10^{-8}
Gly-Phe-Asp-Leu-Ser-Gly	10^{-8}

[a] Relative activity was calculated from the reciprocal of AC_{50} against that of [Ser^5]-speract

Table 5. Respiratory stimulation activity of synthetic [Ser^5]-speract analogs having different amino acid substitution[a]

Sequence	Relative activity
Gly-Phe-Asp-Leu-Ser-Gly-Gly-Gly-Val-Gly	1.0
Gly-Phe-Glu-Leu-Ser-Gly-Gly-Gly-Val-Gly	5.3
Gly-Phe-Asp-Leu-Ser-Gly-Ser-Ser-Val-Gly	0.98
Gly-Tyr-Asp-Leu-Ser-Gly-Gly-Gly-Val-Gly	0.79
Gly-Phe-Asp-Leu-Lys-Gly-Gly-Gly-Val-Gly	0.75
Gly-Phe-Lys-Leu-Ser-Gly-Gly-Gly-Val-Gly	0.35
Gly-Phe-Asp-Leu-Ala-Gly-Gly-Gly-Val-Gly	7.4×10^{-2}
Gly-Phe-Asp-Leu-Ser-Ser-Gly-Gly-Val-Gly	1.7×10^{-2}
Gly-Phe-Asp-Leu-Ser-Gly-Gly-Gly-Ala-Gly	1.6×10^{-2}
Gly-Phe-Ala-Leu-Ser-Gly-Gly-Gly-Val-Gly	5.6×10^{-3}
Gly-Phe-Asp-Leu-Ser-Ser-Ser-Gly-Val-Gly	4.4×10^{-3}
Phe-Gly-Asp-Leu-Ser-Gly-Gly-Gly-Val-Gly	8.7×10^{-4}
Gly-Phe-Asp-Leu-Ser-Gly-Leu-Gly-Val-Gly	2.7×10^{-4}
Gly-Gly-Asp-Leu-Ser-Gly-Gly-Gly-Val-Gly	1.7×10^{-4}
Leu-Ser-Gly-Leu-Ser-Gly-Leu-Ser-Gly-Gly	1.7×10^{-5}
Gly-Phe-Leu-Ser-Gly-Asp-Gly-Gly-Val-Gly	6.8×10^{-6}
Gly-Pro-Asp-Leu-Ser-Gly-Gly-Gly-Val-Gly	3.0×10^{-6}
Gly-Ala-Asp-Leu-Ser-Gly-Gly-Gly-Val-Gly	1.7×10^{-7}

[a] Relative activity was calculated from the reciprocal of AC_{50} against that of [Ser^5]-speract

Table 6. Respiratory stimulation activity of synthetic proline substituted analogs of [Ser⁵]-speract and its D-isomers[a]

Sequence	Relative activity
Gly-Phe-Asp-Leu-Ser-Gly-Gly-Gly-Val-Gly	1.0
Gly-Phe-Asp-Leu-Ser-Gly-Gly-Gly-Val-Pro	500
Gly-Phe-Asp-Leu-D-Ser-Gly-Gly-Gly-Val-Gly	3.8
Pro-Phe-Asp-Leu-Ser-Gly-Gly-Gly-Val-Gly	1.3
Gly-D-Phe-Asp-Leu-Ser-Gly-Gly-Gly-Val-Gly	1.2×10^{-2}
Gly-Phe-Asp-Leu-Ser-Gly-Gly-Gly-D-Val-Gly	3.2×10^{-3}
Gly-Phe-Asp-Leu-Ser-Gly-Gly-Gly-Pro-Gly	2.3×10^{-4}
Gly-Phe-Asp-Leu-Ser-Pro-Gly-Gly-Val-Gly	1.4×10^{-4}
Gly-Phe-Asp-D-Leu-Ser-Gly-Gly-Gly-Val-Gly	1.3×10^{-4}
Gly-D-Phe-D-Asp-D-Leu-D-Ser-Gly-Gly-Gly-D-Val-Gly	3.2×10^{-5}
Gly-Phe-D-Asp-Leu-Ser-Gly-Gly-Gly-Val-Gly	1.4×10^{-5}
Gly-Pro-Asp-Leu-Ser-Gly-Gly-Gly-Val-Gly	3.0×10^{-6}
Gly-Phe-Asp-Leu-Ser-Pro-Gly-Gly-Pro-Gly	1.0×10^{-6}

[a] Relative activity was calculated from the reciprocal of AC_{50} against that of [Ser⁵]-speract

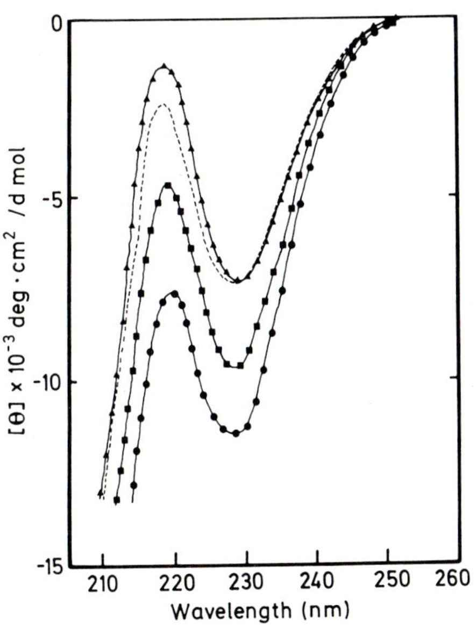

Fig. 5. CD spectra of speract in DDW[a] (----), in 20.8 mM HEPES, (pH 6.8) (–▲–), in 20.8 mM HEPES[b] (pH 6.8) + 12 mM SDS (–■–) and in 20.8 mM HEPES (pH 6.8) + 12 mM SDS + 416.7 mM NaCl (–●–) [a]DDW is *d*eionized, *d*istilled *w*ater;[b] HEPES is N-2-*h*ydrxy-*e*thyl*p*iperazine-N′-2-*e*thane*s*ulfonic acid, [c]SDS is *s*odium *d*odecyl *s*ulfate respiration. Sperm respiration rates were determined at pH 6.8 and the potency of agents was expressed as percentage to sperm respiration rate in normal sea water (pH 8.2)

solution has no ordered structure such as an α-helix or a β-sheet. However, the COOH-terminus which is rich in Gly residues may make a β-turn-like structure under proper conditions. At slightly alkaline pH values or in the presence of 1% SDS the spectra showed a bathochromic shift, suggesting a conformation.

Alloresact contains two cysteine residues. To know whether or not these residues are essential for the respiratory stimulating activity, we chemically synthesized two alloresact analogs ([$Ser^{3,8}$]-alloresact: Lys-Leu-*Ser*-Pro-Gly-Gly-Asn-Ser-Val and *S*-benzyl-alloresact: Lys-Leu-Cys(Bzl)-Pro-Gly-Gly-Asn-Cys(Bzl)-Val) and tested their capabilities for respiratory stimulation. [$Ser^{3,8}$]-alloresact and *S*-benzyl-alloresact stimulated *Glyptocidaris crenularis* sperm respiration at one half of the maximum rate at 2.8 nM and 13 nM, rerspectively [24]. These values are comparable to those of alloresact, thus suggesting that two cysteine residues are not crucial for respiratory stimulation.

4 Biological Activities

4.1 Stimulation of Sperm Respiration and Motility

Decreased respiration rates of sea urchin spermatozoa due to acidification of sea water can be reversed by the addition of sperm-activating peptides (Fig. 1). The stimulated respiration rate did not exceed that of spermatozoa in normal sea water, pH 8.2, even if a large amount of the peptides was added to the sperm suspension medium. This was also true when the respiratory stimulation was induced by solubilized egg jelly. The respiratory stimulation induced by the egg jelly continued for a few minutes and then sperm respiration rates usually declined to the basal rate normally observed at that pH. After reaching the basal state, spermatozoa could again be stimulated by the addition of the peptides (Fig. 1). When the sperm suspending medium was centrifuged before the second addition of the peptides and the resultant supernatant fluid was examined for respiratory stimulating activity toward spermatozoa, the supernatant fluid did not stimulate. This suggests that the peptide binds tightly to spermatozoa.

The activity derived from these peptides has been shown to appear in the ovary in parallel with the progress of ovary maturation [28]. The activity has not present in the immature ovary or the testis.

It is well known that many ionophores cause a change in passive ion transport in resting mitochondria and stimulate mitochondrial respiration [29]. Moreover, it has been reported that a peptide isolated from bovine heart mitochondria induces respiration-dependent Na^+ and K^+ transport in mitochondria [30]. The structure of the peptide is quite similar to that of valinomycin which is a known ionophore for K^+.

Monensin, an ionophore that catalyzes an electro-neutral Na^+/H^+ exchange across cell membranes stimulated sperm respiration and motility, and induced a Na^+-dependent net proton efflux as well as an increased flux of Na^+ in both directions across the cell membrane [27, 31–32]. There seem to be similarities between the effects of egg jelly peptide and monensin. When limited amounts of monensin were used and when the stimulated respiration did not exceed the

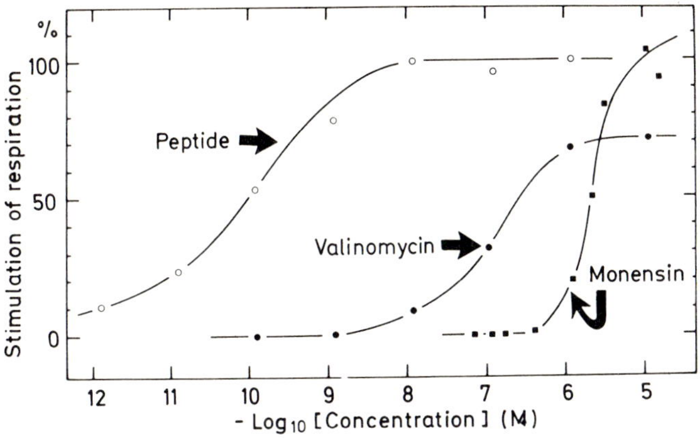

Fig. 6. The potency of synthetic speract (indicated as peptide), valinomycin and monensin in stimulation of *Hemicentrotus pulcherrimus* sperm

respiration in normal sea water, the peptide could stimulate the respiration to the same extent as that obtained in normal sea water. However, the effective concentrations of monensin and valinomycin for respiratory stimulation were much higher than that of the peptide [27]. As shown in Fig. 6, the concentration of the peptide for half-maximal stimulation of sperm respiration was between 10^{-9} M and 10^{-10} M, whereas that of monensin was between 10^{-5} M and 10^{-6} M and that of valinomycin was between 10^{-6} M and 10^{-7} M. The valinomycin-stimulated respiration rate appeared not to reach the rate obtained at pH 8.2. In addition, valinomycin-induced increases in sperm respiration could not be blocked by oligomycin, while monensin- or peptide-induced respiratory stimulation could be blocked by oligomycin. Furthermore, the peptide did not cause any dose-dependent relaxation of the guinea-pig ileum and aorta or of the electrically driven left rabbit atria, whereas monensin was active [27]. This suggests that these organs do not have a building site for the peptide; such a site has been described to be localized in the sperm tail [33].

The respiratory stimulation induced by the peptide and monensin was dependent on the concentration of external Na^+ [27, 31]. Approximately 50 mM Na^+ was required for half-maximal respiratory responses to peptide or monensin (Fig. 7). At concentrations of Na^+ higher than 100 mM, the stimulation reached a plateau.

4.2 Increase of cAMP and cGMP Concentratoins in Sperm Cells

Spermatozoa from various species are known to possess adenylate cyclase, guanylate cyclase, cyclic AMP-dependent protein kinase, cyclic nucleotide phophodiesterase, and phosphoprotein phosphatase activities [34]. In many instances, these enzymes possess higher activity in sperm than in other tissues.

The large molecular weight fucose-sulfate-rich polymer of sea urchin egg jelly causes 100-fold increases in sea urchin sperm cyclic AMP concentrations within

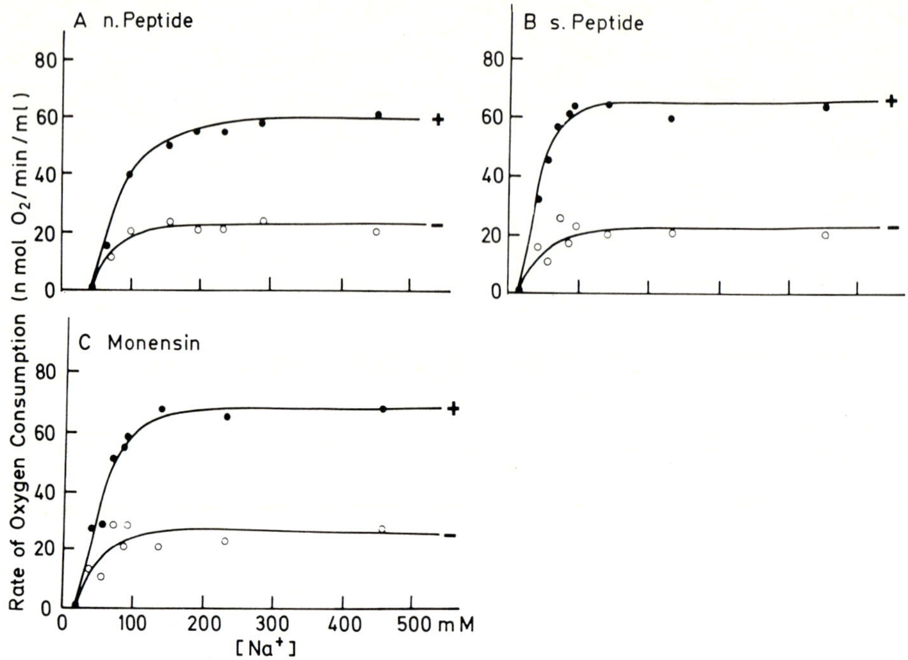

Fig. 7. The effect of extracellular Na^+ concentration on the stimulation of *Hemicentrotus pulcherrimus* sperm respiration by natural (n.peptide) and synthetic (s.peptide) speract or monensin

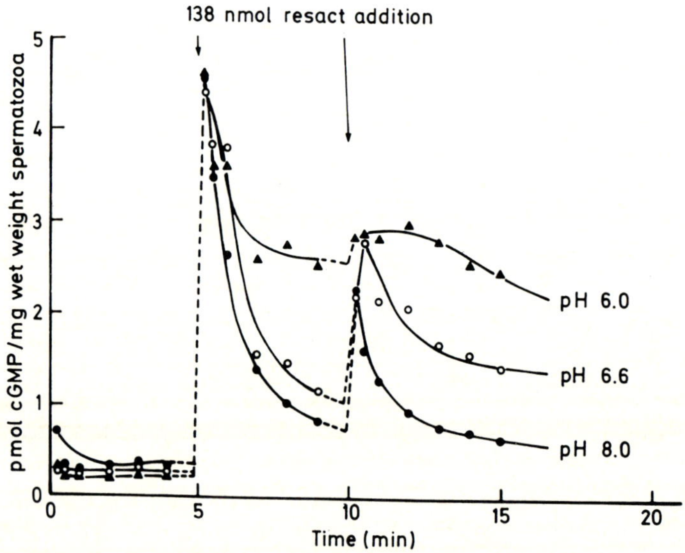

Fig. 8. The effect of resact on cyclic GMP level in *Arbacia punctulata* spermatozoa

1 min, but fails to elevate cyclic GMP [35]. The peptide, however, causes substantial increases in both cyclic AMP and cyclic GMP within a few seconds [17, 19, 24, 31] (Fig. 8). Half-maximal elevations of cyclic GMP are at 2×10^{-9} M and half-maximal elevations of cyclic AMP are at 2×10^{-8} M peptide. Although cyclic GMP concentrations increase slightly in the 10^{-12} to 10^{-11} M range, cyclic AMP concentrations do not appear to change at concentrations of the peptide that markedly stimulate sperm respiration rates. The peptide is an extremely potent activator of sea urchin sperm respiration, but appears somewhat less potent at elevating cylic AMP or cyclic GMP concentrations [13]. The apparent discrepancy between the potency of the peptide toward elevation of cyclic GMP concentrations and the stimulation of sperm respiration may not support the theory that cyclic GMP serves as a second messenger for the peptide.

4.3 Effects on a Major Sperm Plasma Membrane Protein

It has been reported that within seconds after the addition of *Arbacia punctulata* egg jelly to an *Arbacia punctulata* sperm suspension, a 160 KDa sperm protein disappeared and a new protein appeared at 150 KDa [36]. The extent of the change was a function of jelly concentration but, at a given jelly concentration, was independent of incubation time. The change is specific for *Arbacia punctulata* jelly; jelly from other sea urchins such as *Lytechinus pictus*, *Strongylocentrotus purpuratus*, and *S. franciscanus*, at fucose equivalent concentrations 3- to 40-fold greater than those required to induce the complete change with *Arbacia punctulata* jelly, failed to cause any detectable mobility change in the proteins of *Arbacia punctulata* spermatozoa.

Sodium dodecyl sulfate-polyacrylamide gel electrophorsis of the cyanogen bromide cleavage fragments of the 160- and 150-KDa proteins suggests that they are two forms of the same protein [36]. The 160 KDa protein could be labeled with $^{32}PO_4^{3-}$ when intact *Arbacia punctulata* spermatozoa were incubated in sea water containing $^{32}PO_4^{3-}$ and the label was incorporated in the 160 KDa protein in the form [^{32}P]-phosphoserine [36–37]. The label disappeared completely after exposure to jelly, without the corresponding appearance of radioactivity at 150 KDa. When polyclonal antiserum to particulate guanylate cyclase purified from *Strongylocentrotus purpuratus* spermatozoa is used in immnoblots of *Arbacia punctulata* sperm membrane, a single cross-reacting band is observed at 160 KDa [38]. When the spermatozoa are treated with jelly, the cross-reacting band shifts from 160 KDa to 150 KDa. This antiserum also inhibits more than 97% of the guanylate cyclase activity in homogenates of *Arbacia punctulata* spermatozoa. These results suggest that the 160 KDa protein in *Arbacia punctulata* spermatozoa is guanylate cyclase. The factor responsible for the molecular weight change has been identified as a peptide, resact, whose sequence was Cys-Val-Thr-Gly-Ala-Pro-Gly-Cys-Val-Gly-Gly-Arg-Leu-NH_2 [23] (Fig. 9). Speract or mosact also cause a mobility change of a high-molecular-weight sperm membrane protein but it is species specific [16–17, 19]. Monensin and NH_4Cl, two agents known to elevate intracellular pH and to increase sperm respiration rates, also cause a shift in the molecular weight of the membrane protein [36].

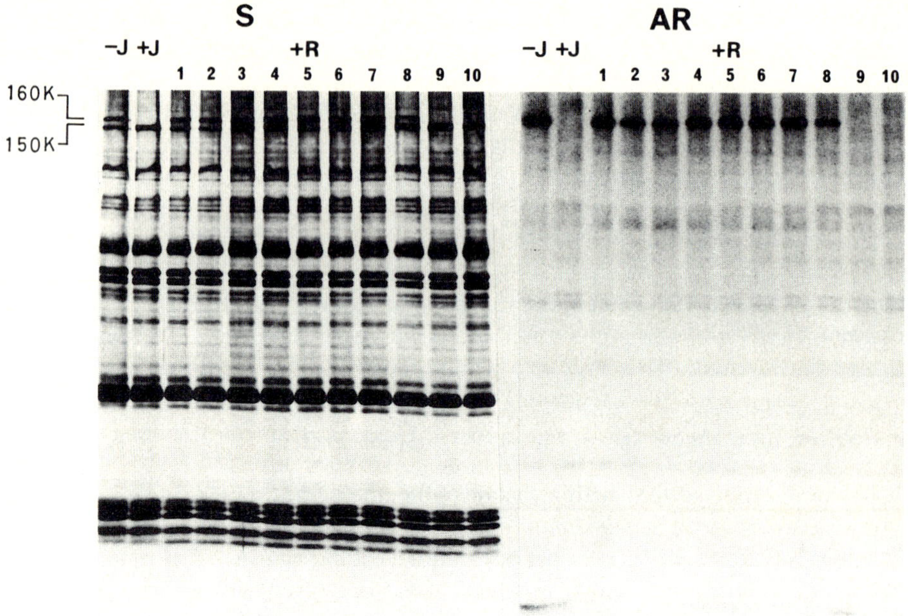

Fig. 9. Dose-dependent modification of an *Arbacia punctulata* sperm protein by resact. Spermatozoa labeled with $^{32}PO_4^{3-}$ were incubated with soluble *Arbacia punctulata* egg jelly or various concentrations of resact. −J, no jelly; +J, with jelly; 1, 1.36×10^{-13} M; 2, 1.36×10^{-12} M; 3, 1.36×10^{-11} M; 4, 1.36×10^{-10} M; 5, 0.68×10^{-9} M; 6, 1.36×10^{-9} M; 7, 0.68×10^{-8} M; 8, 1.36×10^{-8} M; 9, 1.36×10^{-7} M; 10, 1.36×10^{-6} M. At 1 min reactions were stopped by the addition of trichloroacetic acid (final concentration of 10%), and the samples were then centrifuged and the resultant pellets resuspended in ice-cold acetone. After resuspension the samples were again centrifuged and the acetone layer was subsequently removed. The samples were dried and then 10% SDS (Sigma 95% grade material) was added to dissolve the protein and analyzed by slab gel electrophoresis. S, silver-stained gel; AR, autoradiograph of the gel

4.4 Chemoattractant for Sea Urchin Spermatozoa

The chemotaxis of animal spermatozoa to eggs or secretions from the female reproductive system is a widespread phenomenon. Sperm chemotaxis has been demonstrated in four major phyla (Cnidaria, Mollusca, Echinodermata, and Urochordata), and suggestive evidence exists in the Bryozoa, Annelida, Chaetognatha, Nematoda, and the lower vertebrates [39]. However, the chemical nature of animal sperm chemoattractants is unknown although it has been suggested that the attractants of certain echinoderms and hydroids might be peptides [40]. When 10 nM resact was delivered into the *Arbacia punctulata* sperm suspension by the drop by microinjection, many spermatozoa appeared to be forming a cluster in the area of the injected material [41]. As resact diffused from the area of injection, both the number of recruits into the cluster and the size of the cluster continued to grow. The effects of resact on the clustering of spermatozoa were dose-dependent. The clustering of spermatozoa in response to resact required millimolar concentrations of calcium in the surrounding sea water. The actual concentration of resact that elicited a threshold chemotactic response was slightly less than 1 nM. The chemoattractant to resact was species-specific, and the chemotatic response to resact and the dephosphorylation of guanylate cyclase by resact showed similar concentration dependence [41]. Resact is the first egg-derived molecule of known structure shown to be a chemoattractant of animal spermatozoa.

4.5 Acceleration of Sperm Acrosome Reaction

In 1952, Dan first discovered using sea urchin gametes that the acrosome reaction of spermatozoa is essential for fertilization [42]. She described that a responsible factor(s) for the acrosome reaction exists in an extracellular matrix of a sea urchin egg called the jelly layer [42–44]. Since then many investigators have made tremendous efforts to identify the responsible factor(s) and to clarify the mechanism of the reaction [8, 11, 12, 45]. As a result, a fucose sulfate polysaccharide is being considered responsible for induction of the acrosome reaction, and Ca^{2+} entry into spermatozoa is an early step in induction of the acrosome reaction. This influx is followed by ionic changes and increase in the cyclic AMP level. These steps have been considered important for triggering the acrosome reaction. However, the induction of the acrosome reaction by fucose sulfate polysaccharide only occurs fully in Ca^{2+}-enriched sea water. This implies that some other factors are also necessary for induction or promotion of the acrosome reaction under physiological conditions. Ikadai and Hoshi showed that at least two components in starfish egg jelly, a sulfated glycoprotein and a diffusible organic substance, are required to induce the acrosome reaction in starfish spermatozoa [46]. Recently, Matsui *et al.* reported that starfish egg jelly contains a peptide-like substance that activates sperm respiration under slightly acidic conditions, increases the sperm intracellular pH and facilitates the acrosome reaction [47].

As described so far, sea urchin egg jelly also contains oligopeptides that stimulate sperm respiration under slightly acidic conditions, increase the levels of both cyclic AMP and cyclic GMP and induce net hydrogen ion efflux from sperm cells. We then prepared an oligopeptides-free macromolecular component from the egg jelly of the sea urchin *Hemicentrotus pulcherrimus* by gel filtration using normal sea water as an elution solvent and tested the capability to induce the acrosome reaction toward *Hemicentrotus pulcherrimus* spermatozoa at various pHs with or without an oligopeptide, speract [48]. Scanning electron microscopic observation showed that in the absence of speract the macromolecular component induced the acrosome reaction only at half the rate of the component plus speract, or of unfractionated egg jelly at pH 7.4–8.2. The higher rates of induction of the acrosome reaction with speract were associated with great increase of the cyclic AMP concentration in sperm cells. Speract itself, however, did not induce the acrosome reaction in the absence of the macromolecular component of the egg jelly. These results suggest that speract promotes the acrosome reaction as a cofactor, although a major factor(s) responsible for the acrosome reaction is a macromolecular component of the egg jelly.

5 Conclusion

Sea urchin egg jelly consists of two major macromolecular components and several oligopeptides (Fig. 10). These components have important roles for sperm-egg interaction during fertilization. A sialoglycoprotein is responsible for sperm agglutination and a fucose sulfate polysaccharide is a major factor(s) essential for sperm acrosome reaction. Oligopeptides which are specific for spermatozoa of sea urchin species within the same taxonomic order have many biological functions such as stimulation of sperm respiration with Na^+/H^+ exchange across the sperm cell membrane, increase of cyclic AMP and cyclic GMP concentrations in sperm cells and causing a shift in the apparent molecular weight of a major sperm plasma membrane protein which has been proven to be guanylate cyclase with concomitant dephosphorylation of the protein. It has also been known that one of these peptides, resact, acts as a chemoattractant for *Arbacia punctulata* spermatozoa and another, speract, promotes the acrosome reaction as a cofactor. These facts suggest that the peptides in the egg jelly play important roles in the process of normal fertilization of sea urchins.

Most animal eggs have more or less an extracellular matrix like the egg jelly of sea urchins, which probably has the simplest structure. Therefore, it could be imagined that the matrix of other animal eggs including that of mammalian eggs has similar peptides or substances whose function is similar to that of these peptides.

We isolated twenty-three different peptides from nine species of sea urchins distributed over four taxonomic orders. Their structures and activities were specific for sea urchin spermatozoa at the ordinal level. We have given a specific name to each group of peptides. However, we predict that more specific peptides will be isolated from the egg jelly of sea urchins belonging to different orders. Therefore,

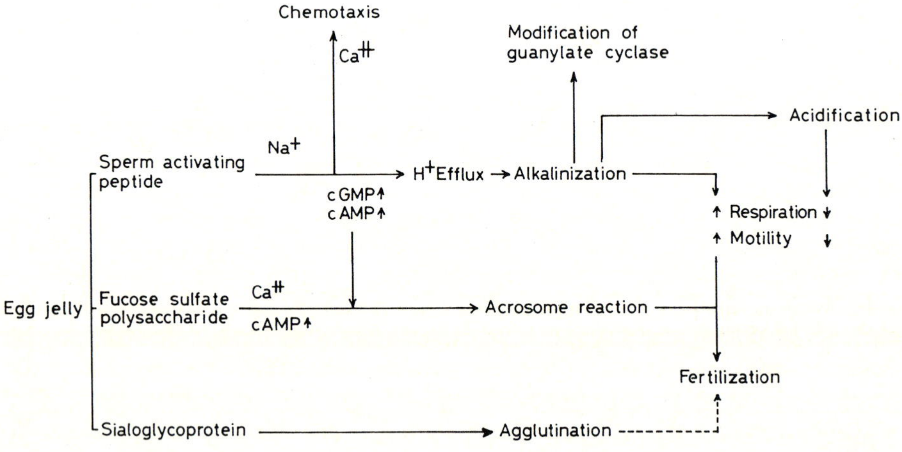

Fig. 10. Schematic presentation of effects of sea urchin egg jelly components on sea urchin spermatozoa

I propose to give these peptides a general name, sperm-activating peptide, abbreviated SAP.

Acknowledgments. The research conducted at Teikyo University School of Medicine and Noto Marine Laboratory was the result of financial support from the Ministry of Education, Science and Culture of Japan from 1982 to 1988 and it is the result of co-work with Dr. Kohji Nomura and Dr. Saburo Isaka. The research conducted at Howard Hughes Medical Institute was collaboration with Professor David L. Garbers. The author thanks Dr. Shyo-ichi Okinaga, Chancellor of Teikyo University, for encouragement to continue the work described here. He also wishes to acknowledge Dr. Masaaki Yamaguchi, two graduate students, Ken-ichi Yoshino and Masanori Kurita, in my laboratory for hard co-working and assistance, and Miss Hiroko Kajiura who did the sequencing with an automated sequencer.

References

1. Lillie FR (1913) J Exp Zool 14:515
2. Hathaway RR (1963) Biol Bull 125:486
3. Ohtake H (1976) J Exp Zool 198:303
4. Ohtake H (1976) J Exp Zool 198:313
5. Kopf GS, Tubb DJ, Garbers DL (1979) J Biol Chem 254:8554
6. Hansbrough JR, Garbers DL (1981) J Biol Chem 256:1447
7. Suzuki N, Nomura K, Ohtake H, Isaka S (1981) Biochem Biophys Res Commun 99:1238
8. Isaka S, Hotta K, Kurokawa M (1970) Exp Cell Res 59:37
9. Hotta K, Hamazaki H, Kurokawa M, Isaka S (1970) J Biol Chem 245:5434
10. Hotta K, Kurokawa M, Isaka S (1970) J Biol Chem 245:6307
11. Ishihara K, Oguri K, Taniguchi H (1973) Biochim Biophys Acta 320:628
12. SeGall GK, Lennarz WJ (1979) Develop Biol 71:33
13. Vasseur E, Hagström M (1946) Ark Zool 37A:1
14. Ciccarelli C, Russo P, De Santis A, Dale B (1987) Develop Growth Differ 29:257
15. Bradley MP, Suzuki N, Garbers DL (1984) Ann N Y Acad Sci 438:142
16. Suzuki N, Kurita M, Yoshino K, Kajiura H, Nomura K, Yamaguchi M (1987) Zool Sci 4:649
17. Suzuki N, Kajiura H, Nomura K, Garbers DL, Yoshino K, Kurita M, Yamaguchi M (1988) Comp Biochem Physiol 89B:687
18. Nomura K, Suzuki N, Ohtake H, Isaka S (1983) Biochem Biophys Res Commun 117:147
19. Shimomura H, Suzuki N, Garbers DL (1986) Peptides 7:491
20. Suzuki N, Hoshi M, Nomura K, Isaka S (1982) Comp Biochem Physiol 72A:489
21. Suzuki N, Yoshino K, Kurita M, Yamaguchi M, Amemiya S (1988) Taxonomical significance of respiratory stimulation of sea urchin spermatozoa by egg associated substances: Proc 6th Internat Echinoderm Conf 23–28 August 1987, Victoria. Available from Balkema AA, Rotterdam
22. Suzuki N, Garbers DL (1984) Biol Reprod 30:1167
23. Suzuki N, Shimomura H, Radany EW, Ramarao CS, Ward GE, Bentley JK, Garbers DL (1984) J Biol Chem 259:14874

24. Suzuki N, Yoshino K, Kurita M, Nomura K, Yamaguchi M (1988) Comp Biochem Physiol 90C:305
25. Yoshino K, Kajiura H, Nomura K, Takao T, Shimonishi Y, Kurita M, Yamaguchi M, Suzuki N (1989) Comp Biochem Physiol (in press)
26. Nomura K, Isaka S (1985) Biochem Biophys Res Commun 126:974
27. Suzuki N, Ohizumi Y, Yasumasu I, Isaka S (1984) Develop Growth Differ 26:17
28. Suzuki N, Kobyashi K, Isaka S (1982) Experientia 38:1245
29. Ovchinnikov YA, Ivanov VT, Shkrob AM (1974) Membrane-active complexones. Elsevier Sci Publ (BBA Library, v 12)
30. Blondin GA, Descastro AF, Senior AE (1971) Biochem Biophys Res Commun 43:28
31. Hansbrough JR, Garbers DL (1981) J Biol Chem 256:2235
32. Repaske DR, Garbers DL (1983) J Biol Chem 258:6025
33. Suzuki N, Kurita M, Yoshino K, Yamaguchi M (1987) Zool Sci 4:641
34. Garbers DL, Kopf GS (1980) The regulation of spermatozoa by calcium and cyclic nucleotides, Raven Press, New York, pp 251–306 (Adv Cyclic Nucleotide Res, v 13)
35. Garbers DL, Hardman JG (1975) Nature 257:677
36. Ward GE, Vacquier VD (1983) Proc Natl Acad Sci USA 80:5578
37. Ward GE, Moy GW, Vacquier VD (1986) J Cell Biol 103:95
38. Ward GE, Garbers DL, Vacquier VD (1985) Science 227:768
39. Miller RL (1985) Sperm chemo-orientation in the metazoa. In Metz CBJr, Monroy A (ed) Biology of fertilization, Academic Press, New York, p 275
40. Miller RL, Tseng CY (1974) Am Zool 14:467
41. Ward GE, Brokaw CJ, Garbers DL, Vacquier VD (1985) J Cell Biol 101:2234
42. Dan JC (1952) Biol Bull 103:54
43. Dan JC (1954) Biol Bull 107:335
44. Dan JC (1956) Int Rev Cytol 5:365
45. Kopf GS, Garbers DL (1980) Biol Reprod 22:1118
46. Ikadai H, Hoshi M (1981) Develop Growth Differ 23:73
47. Matsui T, Nishiyama I, Hino A, Hoshi M (1986) Develop Growth Differ 28:359
48. Yamaguchi M, Niwa T, Kurita M, Suzuki N (1989) Develop Growth Differ 31:233

Cone Shell Toxins and the Mechanisms of Their Pharmacological Action

Masaki Kobayashi, Jun'ichi Kobayashi, and Yasushi Ohizumi[1]

Contents

Abstract

The present review mainly focusses on our recent findings on the structure and mechanism of action of bioactive compounds from cone shells. Cone shells are divided into three groups according to their major prey organisms: worm-eating (vermivorous), shellfish-eating (molluscivorous), and fish-eating (piscivorous). Each species has a venom that reflects the dominant target animals. Various kinds of toxins with novel pharmacological properties have been isolated from cone shell venoms: *Conus geographus* (geographutoxins I and II, α-, μ-, and ω-conotoxins), *C. striatus* (striatoxin), *C. magus* (myotoxin MgTX), *C. eburneus* (eburnetoxin), *C. tessulatus* (tessulatoxin), *C. textile* (arachidonic acid). We also discuss the growing importance of these stinging toxins as tools for studies aimed at elucidating the function of ion channels.

1 Introduction

The cone shells that inhabit tropical and subtropical seas are carnivorous gastropods which catch prey organisms such as fish and shellfish by shooting a venom-containing harpoon-like radular tooth. The venom of one species of cone shell (anboina, *Conus geographus*, Fig. 1) is so potent that it occasionally is responsible

1 Mitsubishi Kasei Institute of Life Sciences, 11 Minamiooya, Machida, Tokyo 194, Japan

Fig. 1. Anboina (*Conus geographus*)

for human fatalities. Thus, the anboina is called "habu-gai" (shellfish that is dangerous like a venomous snake) and is feared in the Okinawan Islands of Japan. On the other hand, cone shells are loved by shell collectors all over the world because of their beautiful colors and shapes. The family Conidae includes approximately 400 species [1], most of which inhabit the Okinawan Islands [2].

Cone shells are divided into three groups according to their major prey organisms. A majority feed on various marine worms such as polychaetes (vermivorous); a smaller number prey on shellfishes (molluscivorous); and a few feed on small bottom fishes (piscivorous). Each species elaborates a venom that reflects the dominant target animals. The venoms of molluscivorous cone shells are highly venomous to gastropods, but are almost ineffective on mice. On the other hand, the venoms of piscivorous Conidae are fatal to both fishes and mice. Further, the symptoms in mice differ depending on the species of cone shell involved. These observations suggest a variety of bioactive principles of cone shell venoms among various species of Conidae.

2 History of Research on Cone Shell Venoms

In the early 1950s, Kohn started ecological studies on cone shells [3], and Endean and coworkers expanded these ecological studies to the toxicology, pharmacology, and biochemistry of Conidae venoms [4, 5]. However, the studies by Endean et al. were performed with crude venoms: thus an important issue has been the purification of bioactive principles of Conidae venoms and the mechanisms of pharmacological action of these compounds.

In 1980, our group started the screening of venoms from about 30 species of cone shells, collected in Okinawa, Japan, by monitoring the pharmacological action on isolated mammalian muscle preparations including skeletal, cardiac, and smooth muscles [6]. In the course of this project, we have isolated numerous bioactive new compounds with a variety of chemical and pharmacological properties from representative species of worm-, shellfish-, and fish-eating cone shells, and have shown that most of these new compounds are peptide toxins acting on ion channels of the cell membrane.

3 What Are Ionic Channels?

Ionic channels are an important site of pharmacological action of natural toxins. It is well known that toxins, even at very low concentrations, block the function of nerves or muscles by stimulating or inhibiting membrane excitability, and thus cause damage to the organism. Ionic channels, small pores that occur in cell membranes and are permeable to ions, directly regulate the excitability of cell membranes [7] (Fig. 2).

Ionic channels consist of transmembrane protein segments and possess two functionally important devices. One of these is the selectivity filter, a narrow region that is responsible for the ionic selectivity among ions, such as Na^+, K^+, Ca^{2+}, and Cl^-, and may be located near the extracellular face of the channel protein. They are referred to as Na channels, Ca channels, etc., depending on their ion selectivity. The other important device is the gate that opens and closes the ionic channel. According to the mode of channel gating, ionic channels are classified into two categories. The first category includes voltage-dependent channels that open and close the gate by monitoring the membrane potential change with a voltage-sensor. Voltage-dependent Na and Ca channels have activation and inactivation gates, while voltage-dependent K channels have only the activation gate. The second

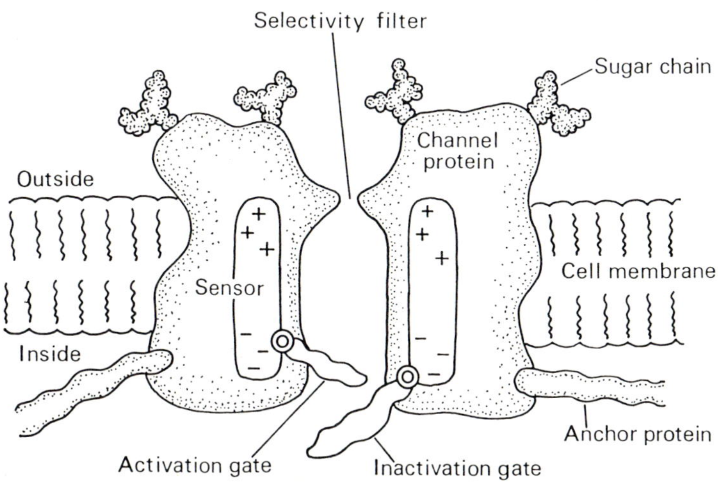

Fig. 2. Model for the ionic channel. The channel is drawn as a transmembrane protein with a hole through the center. The external surface of the molecule is glycosylated. The functional regions, selectivity filter, gate, and voltage sensor are deduced from electrophysiological experiments, but we have yet to elucidate their actual structure

category includes receptor-operated channels that open and close the gate through activation of the receptor by chemical transmitters such as norepinephrine, acetylcholine, and glutamate.

Puffer fish toxin (tetrodotoxin), for example, binds to the entrance part of the Na channel pore to inhibit Na^+ influx and generatoin of an action potential, thus causing the blockade of nerve or muscle function. Aconite toxin (aconitine) or azalea toxin (grayanotoxins) open the activation gate of Na channels, whereas the scorpion or sea anemone toxin (toxin II) inhibits closure of the inactivation gate. These toxins cause subsequent enhancement of Na^+ influx to excite the cells and thus damage nerve or muscle function.

4 Venom Apparatus and Feeding Action of Cone Shells

Among various carnivorous gastropods, cone shells are in a unique position because of the highly specialized radular teeth used as their feeding apparatus. Herbivorous gastropods have hundreds of tiny chitinous radular teeth arranged like in a grater; they shave their algal food with the radular teeth. In the case of cone shells, on the other hand, one centered radular tooth has been developed to become a harpoon. As shown in Fig. 3, the venom apparatus of cone shells consists of a muscular venom bulb, a long coiled venom duct, a radular sheath, radular teeth,

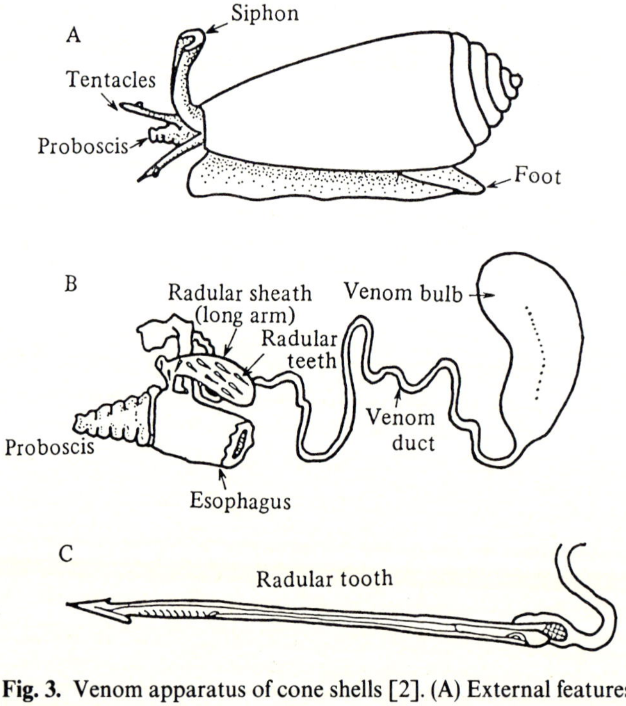

Fig. 3. Venom apparatus of cone shells [2]. (A) External features, (B) venom apparatus, and (C) a radular tooth of a cone shell

and a proboscis [2]. The venom produced in the venom duct is transferred to the radular sheath, where hollow radular teeth are charged with the venom. The bulb contains no venom and is hence thought to provide a pump that transfers the venom from the duct into the radular sheath, which has tens of harpoon-like chitinous radular teeth. The teeth are set one by one along the pharynx to the proboscis, where the foremost tooth is held with its pointed tip forward like a spear. Stinging is accomplished through the thrust of a radular tooth by the proboscis into the prey organism. A tooth is used only once, and if it fails to shoot the prey, that tooth is discarded and a new one from the radular sheath is charged with the venom. The length of radular teeth ranges from several millimeters to one centimeter. The radular tooth has a barb near the tip which prevents it from slipping out once it has taken hold.

5 *Conus geographus*

Conus geographus is called *anboina* as the animal's habitat is the Gulf of Anbon, Indonesia. The shells are about 10 cm in length. *C. geographus* is highly toxic and responsible for human injury. Fatal cases have been reported from Okinawa. Collectors are usually stung while swimming with their catch in a net bag on their side and attempting to examine the shell markings. Stings by *C. geographus* immediately produce severe pain and numbness at the injured area. Then numbness spreads around the mouth and to the extremities. In most cases, this is accompanied by vomiting, dizziness, lacrimation, hypersalivation, and chest pain. In severe cases dyspnea, dysphagia, aphonia, blurring of vision, motor incoordination, generalized pruritus, respiratory paralysis, and death follow [2].

It has been reported [5, 6] that the crude venom of *C. geographus* causes an inhibitory effect on the contractile response of stimulated skeletal muscle.

We successfully isolated two peptide toxins composed of 22 amino acid residues including 3-hydroxyprolines by monitoring the pharmacological activity [8]. As shown in Fig. 4, the primary structures of the two peptides named geographutoxin I and II (GTX I and II) are very similar to each other [9]. They differ by four amino

5 10 15

Geographutoxin I: Arg-Asp-Cys-Cys-Thr-Hyp-Hyp-Lys-Lys-Cys-Lys-Asp-Arg-Gln-Cys-
(GTX I)

20

Lys-Hyp-Gln-Arg-Cys-Cys-Ala-NH_2

5 10 15

Geographutoxin II: Arg-Asp-Cys-Cys-Thr-Hyp-Hyp-Arg-Lys-Cys-Lys-Asp-Arg-Arg-Cys-
(GTX II)

20

Lys-Hyp-Met-Lys-Cys-Cys-Ala-NH_2

Fig. 4. Amino acid sequences of geographutoxin I and II

acid residues, at positions 8, 14, 18, and 19 from the C terminal. The substitution of the amino acids was explained by single base substitution in the triplet codons except for the substitution at residue 18 of Gln in GTX I by Met in GTX II. Distinctive features of the primary structures are a basic amino acid cluster in the middle of the sequence, and a Cys-Cys sequence in both N and C terminal regions. The LD_{50} values (i.p.) of GTX I and II in mice are 340 $\mu g\,kg^{-1}$ and 110 $\mu g\,kg^{-1}$, respectively.

GTX II at very low concentrations inhibits twitch responses of skeletal muscle [10,11]. The IC_{50} value of GTX II (3×10^{-8} M) is somewhat higher than that of tetrodotoxin (1×10^{-8} M) (Fig. 5). GTX II inhibited action potentials of skeletal muscle without change in the resting potential. Surprisingly, action potentials of the crayfish giant axon, neuroblastoma N1E-115 cells, and cardiac muscle were not affected by GTX II even at high concentrations [12] (Fig. 6).

Toxin-binding studies have shown that tetrodotoxin and saxitoxin bind at a common receptor site on sodium channels, which has been designated neurotoxin

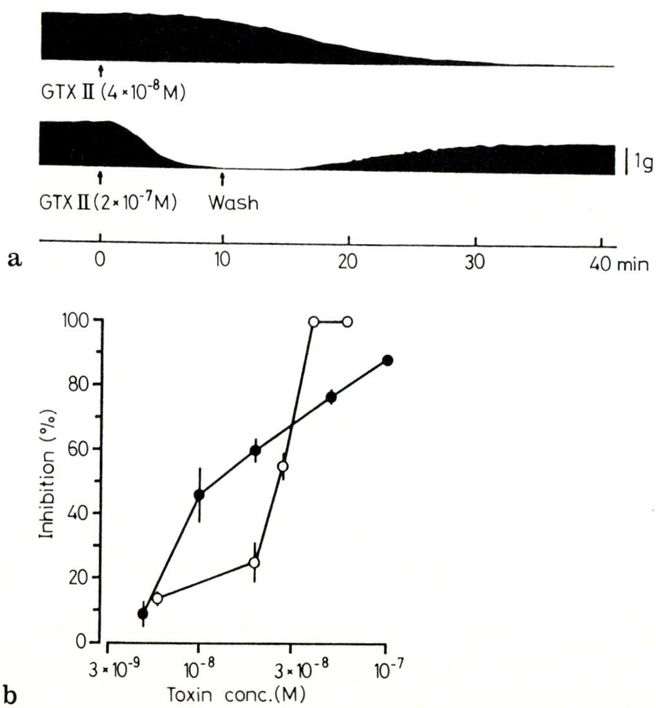

Fig. 5a, b. Inhibitory effects of geographutoxin II (GTX II) on the contractile response of the isolated mouse hemidiaphragm to direct electrical stimulation (0.1 Hz, 5 msec, supramaximal voltage). **a**, typical inhibition patterns of twitch responses by GTX II and its recovery after washing out the toxin. **b**, dose-inhibitory response curve for GTX II (o) or TTX (•). The tensions of twitch responses were measured 60 min after exposure to the toxins. Each point and bar represent the mean ±S.E.M. (n=3–5). Conc, concentration

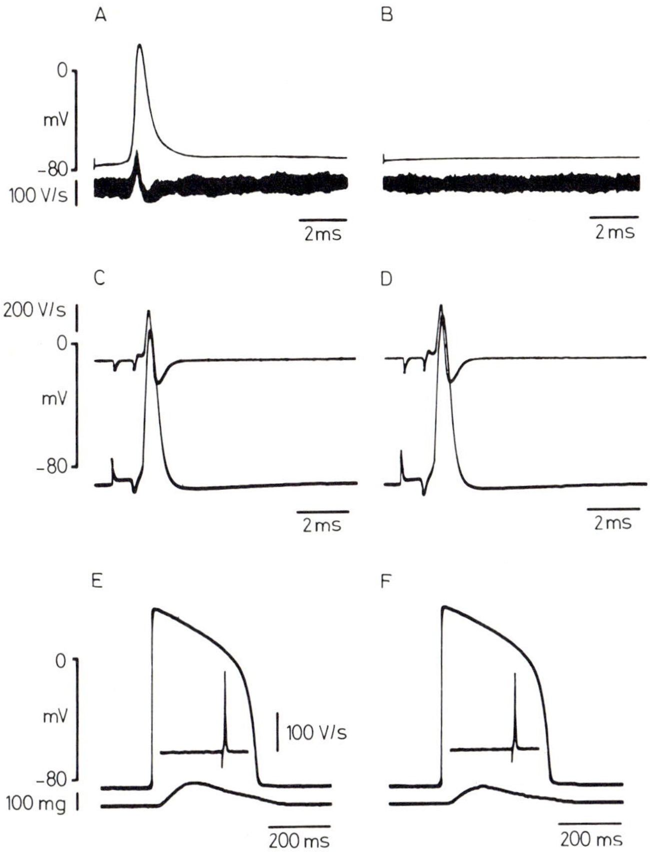

Fig. 6. Effects of geographutoxin II (GTX II) on various tissues. A and B, effect on the action potential of guinea-pig extensor digitorum longus muscle. Action potential (upper trace) and its derivative (lower trace) before (A) and 6 min after (B) application of 5×10^{-7} M GTX II. Ca^{2+}-free medium was used to depress muscle contraction. C and D, effect on crayfish giant axon. Action potential (lower trace) and its derivative (upper trace) before (C) and 1 hr after (D) application of 5×10^{-6} M. E and F, effect on guinea-pig papillary muscle of the heart. Action potential (upper trace) and contraction (lower trace) before (E) and 30 min after (F) application of 1×10^{-6} M GTX II. The derivative of the corresponding action potential shown in the inset was recorded at twice faster sweep speed

receptor site 1. GTX II inhibits [^{3}H]saxitoxin binding to Na channels in skeletal muscle homogenates and T-tubular membranes in close agreement with concentrations that block muscle contraction [13]. However, [^{3}H]saxitoxin binding to membranes of superior cervical ganglion is not affected by GTX II even at high concentrations. Scatchard analysis of [^{3}H]saxitoxin binding to T-tubular membranes revealed a primarily competitive mode of inhibition of saxitoxin binding by GTX II (Fig. 7). These results indicate that GTX II interacts competitively with saxitoxin in binding at neurotoxin receptor site 1 of the Na channel in a highly

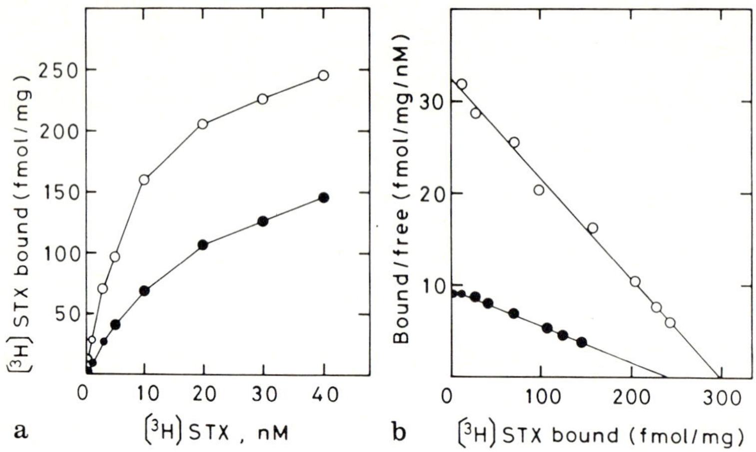

Fig. 7a, b. Binding of [^{3}H]saxitoxin ([^{3}H]STX) to T-tubular membranes of rabbit skeletal muscles in the presence or absence of GTX II. T-tubular membranes were incubated with increasing concentrations of [^{3}H]STX from 0.4 to 40 nM for 30 min at 36° C and bound STX was measured using the rapid membrane filtration assay. Nonspecific binding in the presence of 2 μM tetrodotoxin has been subtracted from the results. In **a**, [^{3}H]STX binding was measured in the presence (●) or absence (○) of 50 nM GTX II and are plotted. In **b**, [^{3}H]STX binding in the presence (●) or absence (○) of GTX II is presented as a Scatchard plot

tissue-specific manner. Previously no differences had been found experimentally, between muscle and nerve Na channels because of their electrophysiological and pharmacological similarity [14]. Thus, GTX II is the first reagent that discriminates at this site between nerve and muscle Na channels. GTX II is used to identify these tissue-specific regions of the Na channel structure at neurotoxin receptor site 1 [15].

Olivera et al. recently isolated from *C. geographus* three classes of peptide toxins named α-, μ-, and ω-conotoxins [16] (Table 1). α-Conotoxin GI, GIA, GII, and MI are composed of 13–15 amino acid residues including two S-S bridges per molecule and are homologous with one another. These peptides cause postsynaptic inhibition at the neuromuscular junction and inhibit competitively α-bungarotoxin and curare alkaloid binding to nicotinic acetylcholine receptors. Many pharmacological studies have revealed that α-conotoxins are nicotinic receptor blockers. μ-Conotoxin GVIIIA, GVIIIB, and GVIIIC with 22 amino acid residues are rich in hydroxyproline. These peptide toxins block Na channels in muscle much more effectively than those in nerve. The peptides which our group have named GTX I and II may be identical with GVIIIA and GVIIIB.

ω-Conotoxins GVIA, GVIB, GVIC, GVIIA, and GVIIB called "shaker peptides", were also isolated from *C. geographus*. These peptides induce a persistent tremor when injected intracerebrally in mice. Amino acid sequences (25–29 amino acid residues) of ω-conotoxins are shown in Table 1. The transmitter release at the neuromuscular junction of skeletal muscle was blocked by ω-conotoxin GVIA,

Table 1. Amino acid sequences of conotoxin [16]

α	GI	E C C	N P	A	C G R	H Y S C[a]						
	GIA	E C C	N P	A	C G R	H Y S C	G K[a]					
	GII	E C C	H P	A	C G K	H F S C[a]						
	MI	G R C C	H P	A	C G K	N Y S C[a]						
μ	GVIIIA	R D C C	T Hy	Hy	K K C	K D R Q	C K	Hy	Q R C C A[a]			
	GVIIIB	R D C C	T Hy	Hy	R K C	K D R R	C K	Hy	M K C C A[a]			
	GVIIIC	R D C C	T Hy	Hy	K K C	K D R R	C K	Hy	L K C C A			
ω-"Shaker-Blocker"	GVIA	C K S Hy	G S	S	C S Hy	T S Y N	C C	R	– S C N Hy	Y T K R C	– – Y[a]	
	GVIB	C K S Hy	G S	S	C S Hy	T S Y N	C C	R	– S C N Hy	Y T K R C	– – Y	G
	GVIC	C K S Hy	G S	S	C S Hy	T S Y N	C C	R	– S C N Hy	Y T K R C		
ω-"Shaker"	GVIIA	C K S Hy	G T	Hy	C S R	G M R D	C C	T	– S C L L	Y S N K C	R R Y	
	GVIIB	C K S Hy	G T	Hy	C S R	G M R D	C C	T	– S C L S	Y S N K C	R R Y	
	MVIIA	C K G K	G A	K	C S R	L M Y D	C C	T	G S C R –	– S G K C[a]		
"Sleeper"	GV	G E γ γ	L Q	γ	N Q γ	L I R γ	K S	N[a]				

[a] The α-carboxyl is known to be amidated; absence of [a] indicates that no assignment has yet been made. A (Ala), C (Cys), D (Asp), E (Glu), F (Phe), G (Gly), H (His), K (Lys), L (Leu), N (Asn), P (Pro), Q (Glu), R (Arg), S (Ser), T (Thr), Y (Tyr), γ (γ—carboxyglutamate), Hy(*trans*-4-hydroxyproline).

GVIB, and GVIC, whereas it was not affected by other ω-conotoxins (GVIIA and GVIIB). Electrophysiological experiments indicate that ω-conotoxin GVIIA acts by blocking Ca channels without affecting the propagation of actoin potentials in the nerve terminal.[17]. Voltage-dependent Ca channels are divided into 3 classes, L-, N-, and T-type Ca channels, on the basis of electrophysiological and pharmacological properties. The toxin blocks selectively L- and N-type Ca channels in neurons but not L- and T-types in muscle [18]. Ca antagonists (1,4-dihydropyridine derivatives) have been extensively used for studying Ca channels as a pharmacological or biochemical probe, but they cannot discriminate between T- and N-type Ca channels. Therefore, ω-conotoxins may provide a valuable tool for defining structural differences among these Ca channel subtypes in muscle and nerve [19]. In addition, Olivera et al. isolated a sleeper peptide (GV) from the same shell. In mice, GV causes a sleeplike state when injected intracerebrally.

6 *Conus striatus*

Conus striatus, a piscivorous member of the Conidae, is not sufficiently venomous to be of potential danger to humans. But, several severe cases of poisoning by *C. striatus* have been reported. We have isolated from this venom a glycoprotein (molecular weight 25,000) named striatoxin (StTX) as a cardiotonic component [20].

Application of StTX to the isolated guinea-pig left atria caused a concentration-dependent inotropic effect; the effect of StTX was reversed in the presence of tetrodotoxin (Fig. 8). In the atria, StTX provoked action potentials with a plateau phase of long duration without affecting the maximum rate of rise and amplitude of the action potential and the resting membrane potential. This prolongation was also reversed by tetrodotoxin. These results suggest that StTX causes prolongation of the action potential duration probably due to slowed inactivation of Na inward

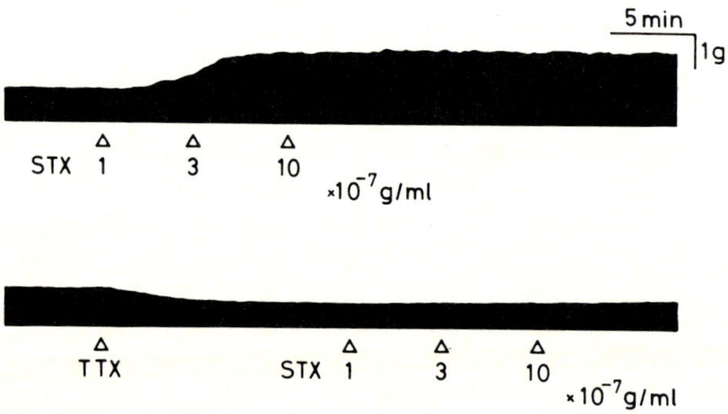

Fig. 8. Effects of striatoxin (StTX) on guinea-pig isolated left atria in the presence (lower trace) or absence (upper trace) of tetrodotoxin (TTX, 10^{-6} M). The atria were driven by electrical stimulation at frequency of 2 Hz with square-wave pulses of 5 msec at 5 V. TTX was added 15 min before administration of StTX

currents and that this may result in an increase in Ca^{2+} availability in cardiac muscle cells. This could explain the cardiotonic action of StTX. Furthermore, in mouse neuroblastoma cells StTX slowed Na channel inactivation without affecting the time course of activation of the channels, and the voltage dependence of activation was shifted to more negative membrane potentials. The binding of saxitoxin or α-scorpion toxin to Na channels was not affected by StTX, while that of batrachotoxin was slightly enhanced by the toxin. It may be concluded that StTX interacts with a new receptor site on Na channels at which specific effects on channel inactivation can occur [21]. In the isolated guinea-pig ileum, StTX caused rhythmic transient contraction followed by relaxation, which was reversed by tetrodotoxin and a low-Na^+ medium. Our pharmacological studies on StTX suggest that the contraction of the ileum induced by the toxin is due to the excitation of cholinergic nerves, while the relaxation is mediated through non-adrenergic inhibitory nerves [22]. Substances possessing both excitatory and inhibitory activities, such as StTX, had not been previously reported except the anthopleurins, cardiotonic peptides (molecular weight approximately 5,000) isolated from sea anemones [23]. Incidentally, the minimum lethal dose of StTX in a fish, *Rhodeus ocellatus smithi*, is 1 $\mu g\,g^{-1}$ body weight.

7 *Conus magus*

The venom of *C. magus* is lethal to fish and mice, as are those of *C. geographus* and *C. striatus*, although the average size of *C. magus* is only about 5 cm long and smaller than *C. geographus* and *C. striatus*. From the venom of *C. magus* was isolated a myotoxin (MgTX), a peptide toxin with an estimated molecular weight of 1,500 by gel filtration [24]. MgTX elicits an increase in contractile force of the isolated guinea-pig ileum. The mode of action of MgTX is similar to that of the cardiotonic glycoprotein StTX isolated from *C. striatus*. Another active component (MAC) in the venom of *C. magus* was also purified and characterized as a protein of molecular weight between 45,000 and 65,000 [25]. MAC exhibits rhythmic transient contractions of the guinea-pig ileum followed by relaxations and a marked increase in the contractile force of the guinea-pig left atria, as in the case of StTX.

On the other hand, two peptide toxins, MI and MVIIA were isolated from the Philippine *C. magus* by Olivera et al. [16]. The chemical and pharmacological properties of MI and MVIIA are different from those of MAC and MgTX isolated from the Okinawan *C. magus*, but resemble those of GII and GVIIA isolated from the Philippine *C. geographus*. MI and MVIIA are classified as α- and ω-conotoxins, respectively. MI showed no rhythmic contractions of the guinea-pig ileum, suggesting that the pharmacological properties of MgTX and MI differ from each other.

8 *Conus eburneus*

C. eburneus is a vermivorous member of the Conidae, which feeds on polychaete worms. The average size of the shell is about 6 cm long and they inhabit sandy seashores. From the venom of the Okinawan *C. eburneus* was purified a vasoactive protein (molecular weight 28,000) named eburnetoxin (ETX) [26]. ETX induces contraction of isolated rabbit aorta, resulting from a remarkable increase in Ca^{2+} permeability in the cell membrane.

9 *Conus tessulatus*

C. tessulatus is also a vermivorous member of the Conidae as is *C. eburneus*; the average size of the shell is approximately 5 cm long. A vasoactive protein (molecular weight 26,000) designated tessulatoxin (TsTX), was isolated from the venom of the Okinawan *C. tessulatus* [27]. The pharmacological actoin of TsTX is quite similar to that of ETX, the vasoactive substance from *C. eburneus*.

Recently, Lazdunski et al. isolated a protein toxin consisting of two subunits (molecular weight 26,000 and 29,000) from the venom of the Madagascan *C. tessulatus* [28]. This protein toxin activates the Na^{+}–Ca^{2+} exchange system by increasing Na^{+} permeability of the cell membrane to result in increased Ca^{2+} permeability of cardiac muscle cells, or to induce contraction of smooth muscle.

10 *Conus textile*

C. textile is a molluscivorous member of the Conidae and its prey organisms are closely related molluscivorous Conidae. The average size of the shell is comparable to that of *C. geographus*. Human fatalities by the venomous sting of *C. textile* are recorded in some dated articles. We examined the toxicity of the Okinawan *C. textile* venom against mice, but found that no mouse died even after intraperitoneal injection of a large amount of the venom. The human fatalities recorded in the literature, therefore, may result from the venomous sting of *C. geographus* but not from that of *C. textile*. Arachidonic acid was isolated from the venom of *C. textile* as a principle of the powerful transient contraction of the isolated guinea-pig ileum. The amount of arachidonic acid in the contents of the venom duct was estimated to be as much as 6 $mg\,g^{-1}$ wet weight (0.6%) [29, 30], but the role of arachidonic acid in the venom remains uncertain.

11 Food Habits of Cone Shells and Pharmacological Properties of the Venoms

It has been asserted that the venoms of piscivorous Conidae are lethal to mice as well as to fish, while those of molluscivorous and vermivorous Conidae are toxic to mollusks (gastropods, etc.) and worms (polychaete worms, etc.), which are their prey organisms, but not toxic to mice. As described in the present article, the peptide toxins isolated from piscivorous cone shells such as *C. geographus*, *C. striatus*, and *C. magus*, potently paralyze the mouse skeletal muscle, whereas the toxic principles of the molluscivorous *C. textile* and the vermivorous *C. eburneus* and *C. tessulatus* have no effect on skeletal muscle, but induce marked excitation of

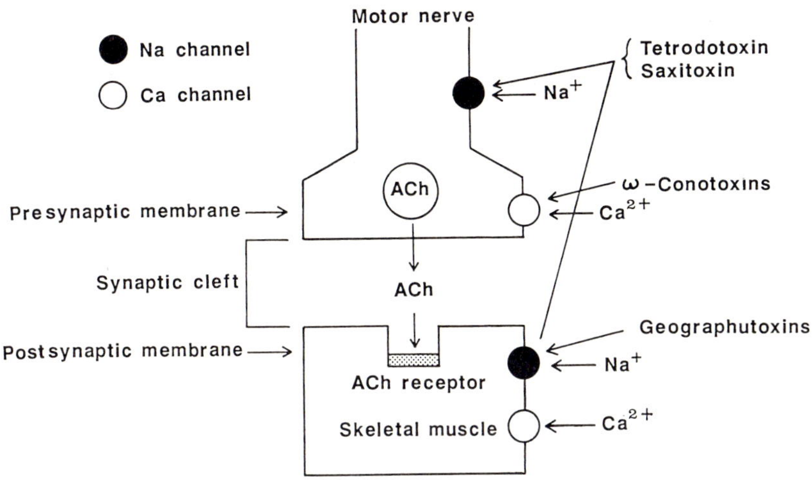

Fig. 9. Sites of action of cone shell toxins, tetrodotoxin, and saxitoxin in the neuromuscular junctional region

smooth muscles (aorta, ileum, etc.). Thus, our pharmacological studies corroborate the belief that each kind of cone shell possesses its own venom which is effectively toxic to its prey organisms.

Recently, a Na channel protein with molecular weight of 260,000 was purified by using tetrodotoxin, which binds specifically to the Na channel [31]; the amino acid sequence of the Na channel protein was determined by genetic engineering techniques. As described above, the venom of *C. geographus* contains several types of peptide toxins which act through different mechanisms from one another, namely α-conotoxin which blocks nicotinic acetylcholine receptors, ω-conotoxin which blocks Ca channels, and geographutoxin (μ-conotoxin) which blocks Na channels (Fig. 9). Among these toxins, geographutoxin (GTX II) and ω-conotoxin (GVIA) are expected to act as chemical probes to classify Na channel subtypes and as tools to clarify the molecular mechanism of nerve Ca channels, by physiologists, pharmacologists, biochemists, and biologists all over the world. Thus, the cone shell toxins will be increasingly used as valuable tools in studies aimed at elucidating the molecular mechanisms of various complicated cellular events in living organisms.

References

1. Halstead BW (1978) Poisonous and venomous marine animals of the world, rev ed, Darwin press, New Jersey, p 184
2. Hashimoto Y (1979) Marine toxins and other bioactive marine metabolites, Japan Scientific Societies Press, Tokyo, p 188
3. Kohn AJ (1959) Ecol Monogr 29:47
4. Endean R, Rudkin C (1965) Toxicon 2:225

5. Endean R, Parish G, Gyr P (1974) Toxicon 12:131
6. Kobayashi J, Nakamura H, Hirata Y, Ohizumi Y (1982) Toxicon 20:823
7. Ohizumi Y, Kobayashi J, Nakamura H (1986) Chemistry Today 179:14
8. Nakamura H, Kobayashi J, Ohizumi Y, Hirata Y (1983) Experientia 39:590
9. Sato S, Nakamura H, Ohizumi Y, Kobayashi J, Hirata Y (1983) FEBS Lett 155:277
10. Minoshima S, Kobayashi M, Nakamura H, Kobayashi J, Ogura A, Takahashi M, Ohizumi Y (1984) Jpn J Pharmacol 36:1920
11. Ohizumi Y, Minoshima S, Takahashi M, Kajiwara A, Nakamura H, Kobayashi J (1986) J Pharmacol Exp Ther 239:243
12. Kobayashi M, Wu CH, Yoshii M, Narahashi T, Nakamura H, Kobayashi J, Ohizumi Y (1986) Pflügers Arch 407:241
13. Ohizumi Y, Nakamura H, Kobayashi J, Catterall WA (1986) J Biol Chem 261:6149
14. Campbell DT, Hille B (1976) J Gen Physiol 67:309
15. Gonoi T, Ohizumi Y, Nakamura H, Kobayashi J, Catterall WA (1987) J Neurosci 7:1728
16. Olivera BM, Gray WR, Zeikus R, McIntosh, Varga J, Rivier J, de Santos V, Cruz LJ (1985) Science 230:1338
17. Kerr LM, Yoshikami D (1984) Nature 308:282
18. McCleskey EW, Fox AP, Feldman D, Tsien RW (1986) J Exp Biol 124:177
19. Cruz LJ, Olivera BM (1986) J Biol Chem 261:6230
20. Kobayashi J, Nakamura H, Hirata Y, Ohizumi Y (1982) Biochem Biophys Res Commun 105:1389
21. Gonoi T, Ohizumi Y, Kobayashi J, Nakamura H, Catterall WA (1987) Mol Pharmacol 32:691
22. Kobayashi J, Nakamura H, Ohizumi Y (1981) Br J Pharmacol 73:583
23. Ohizumi Y, Shibata S (1981) Br J Pharmacol 72:239
24. Kobayashi J, Nakamura H, Ohizumi Y (1983) Eur J Pharmacol 86:283
25. Kobayashi J, Nakamura H, Ohizumi Y (1985) Toxicon 23:783
26. Kobayashi J, Nakamura H, Hirata Y, Ohizumi Y (1982) Life Sci 31:1085
27. Kobayashi J, Nakamura H, Hirata Y, Ohizumi Y (1983) Comp Biochem Physiol 74:381
28. Schweitz H, Renaud JF, Randimbivololona N, Preau C, Schmid A, Romey G, Rakotovao L, Lazdunski M (1986) Eur J Biochem 161:787
29. Kobayashi J, Ohizumi Y, Nakamura H, Hirata Y (1981) Toxicon 19:757
30. Nakamura H, Kobayashi J, Ohizumi Y, Hirata Y (1982) Experientia 38:897
31. Noda M, Shimizu S, Tanabe T, Takai T, Kayano T, Ikeda T, Takahashi H, Nakayama H, Kanaoka Y, Minamino N, Kangawa K, Matsuo H, Raftery MA, Hirose T, Inayama S, Hayashida H, Miyata T, Numa S (1984) Nature 312:121

Epibiosis of Marine Algae and Benthic Invertebrates: Natural Products Chemistry and Other Mechanisms Inhibiting Settlement and Overgrowth

Andrew R. Davis[1,4], Nancy M. Targett[2], Oliver J. McConnell[3], and Craig M. Young[1]

Contents

1 Division of Marine Science, Harbor Branch Oceanographic Institution, Fort Pierce, FL 34946, USA

2 College of Marine Studies, University of Delaware, Lewes, DE 19958, USA

3 Division of Biomedical Marine Research, Harbor Branch Oceanographic Institution, Fort Pierce, FL 34946, USA

4 Current Address: Department of Biology, University of Wollongong, GPO Box 1144 Wollongong, N.S.W 2500 Australia

Bioorganic Marine Chemistry, Volume 3

Abstract

This article reviews chemical and physical mechanisms by which non-motile marine invertebrates and marine algae inhibit the fouling of their surfaces. Inhibition of primary film formation (bacterial, diatoms and protozoans); and macrophyte and epizooite settlement are considered individually followed by discussion of their interactions. Emphasis is placed on ecological and evolutionary considerations, including the advantages and disadvantages for an organism in being fouled, variation in the defenses utilized by an organism and the likelihood of misinterpreting interactions among organisms. The study of natural antifouling defenses is an emerging field offering important theoretical and economic spin-offs.

1 Introduction

Ecology may be considered as the study of interactions among organisms and between organisms and their environment. Examination of interactions in terrestrial ecosystems, and more recently in marine ecosystems has revealed a significant role of chemical compounds synthesized by organisms (natural products) in mediating many of these interactions [1–8]. Natural products may function as pheromones, feeding deterrents, homing and site recognition cues, and fouling inhibitors, to name just a few examples [1, 9, 10]. A large number of natural products have been isolated from marine organisms [11–17] but the functions (if any) of many of these compounds remain largely unknown.

This review discusses the settlement of bacteria, diatoms, protozoa, algal spores, and invertebrate larvae from the point of view of a biotic substratum. More specifically, we will review the role of marine natural products and physical mechanisms in mediating the inhibition of settlement onto, and overgrowth of, living substrate (sessile benthic invertebrates and algae).

1.1 Colonization of a Surface

On immersing a surface in seawater, a complex process of colonization is initiated [18–22]. The surface is initially 'conditioned' by a glycoproteinaceous film which is adsorbed almost instantly following immersion [20]. Bacteria, diatoms, and protozoa (the primary film) then colonize the surface, usually followed by the attachment of macrophytes and epizooites [21, 23–25]. These three groupings (the primary film, macrophytes and epizooites) represent convenient and logical foci for discussion. The mechanisms which organisms use to inhibit each of these groups of settlers will be considered in turn. The separation of organisms into these three arbitrary groups belies the potentially important interactions which may occur between them. We therefore devote a section to the consideration of these interactions.

Surfaces may be colonized by water-borne adults or propagules (i.e., spores or larvae), or alternately by the lateral overgrowth of neighboring organisms [26]. Some nonmotile invertebrates possess crawling larvae [27, 28], but free-swimming larvae are more common.

Organisms within each group may colonize surfaces at different stages in their life histories and in subtly different ways. Differences in the colonization process may be important considerations in the development of mechanisms to inhibit the attachment of the fouling organisms. For example, components of the primary film colonize surfaces as adults, and presumably do so throughout the year. Macrophytes and epizooites, on the other hand, usually colonize substrata at early stages in their life histories and therefore colonization is restricted to specific reproductive seasons, at least in temperate latitudes. Overgrowth of a surface by neighboring organisms may occur during any season [29], although there may be clear seasonal differences in the rates of overgrowth, especially in temperate climates.

Organisms in the three groupings vary in their ability to distinguish among substrata. For bacteria, the initial attachment is by simple stickiness due to adhesives, electrostatic forces, electrical double layer phenomena, or London/Van der Waals forces [30, 31]. Chemotaxis may be an important mechanism in the initial attraction (or repulsion) of a microorganism to a surface. Motile bacteria in the genera *Bacillus*, *Pseudomonas*, *Vibrio*, *Escherichia*, and *Salmonella* are capable of moving along chemical gradients [32, 33]. Some gliding bacteria, fungi, and dinoflagellates also respond to chemical stimuli [32, 33].

Algae differ in their ability to select substrata. In green and brown algae, the asexual and motile zoospores or swarmers are primarily involved in substratum colonization, whereas red algae utilize nonmotile carpospores and tetraspores. During the first few hours following release from their respective sporangia, the motile zoospores usually swim towards areas of low light intensity and roughened surfaces. In contrast, the nonmotile spores of red algae move with prevailing water currents, while sinking slowly onto substrates. Consequently, the red algal carpospores and tetraspores are incapable of site selection. Motile brown and green algal zoospores utilize flagella for site selection and initial adherance. The flagella are subsequently withdrawn into the respective cells or are shed. In red algal zoospores, a characteristic mucilagenous, polysaccharide sheath is probably involved in initial attachment to substrata. After initial settlement, all algal spores release an adhesive material which strengthens the substratum attachment. This material is believed to be composed of a polysaccharide-protein complex. An outer cell wall layer is then secreted immediately by the algal spores, probably to protect the spores from unfavorable conditions [34].

The larvae of many marine invertebrates may respond to a plethora of environmental cues, with a combination of light, tactile and chemosensory organs [35]. Hence many invertebrate larvae are highly discriminatory and may be capable of locating and settling on substrata which are likely to improve their subsequent survivorship [36–42].

1.2 Mechanisms of Inhibition

Artificial substrata, such as wood, ceramic tiles, and unprotected metals, rapidly develop a covering of epifaunal organisms [23]. Organisms which possess hard surfaces, such as bivalve shells or the tests of some solitary ascidians, would also appear to provide excellent settlement sites for fouling organisms [43, 44]. Indeed, at the spatial scales at which a larva encounters substrata, even flexible or porous organisms such as fleshy algae and sponges may potentially provide sites for settling organisms. Yet many organisms are essentially unfouled, and furthermore, the inhibition of fouling is usually observed only among living organisms [but see 45]. For example, dead gorgonians are rapidly fouled by a variety of invertebrates [46].

If a surface (i.e., organism) is to inhibit the settlement of epibionts, it may act in either of the following physical or chemical ways:

a) Deter potential settlers in the water column before they have encountered the surface [47–49].
b) Deter potential settlers on their encounter with the surface by offering negative cues to propagules [50–53] or poison juveniles after their settlement [54].

1.3 Economic and Environmental Considerations

The control of biofouling is of particular concern to modern marine engineering and shipping operations. Uncontrolled settlement and adhesion of marine invertebrates to the hulls of ships increase frictional drag, with a corresponding decrease in speed, maneuverability, and fuel efficiency. Triorganotin compounds, chiefly tributyltin, are effective inhibitors of settlement, but may be toxic to many nontarget species of commercial importance [55–58]. Deleterious effects of tributyltin have been observed at 10 ng/L; this compound may attain many times this concentration in some marinas [59]. The application of nontoxic marine natural products which are capable of inhibiting the fouling of ships and other structures may potentially provide a solution to this sensitive environmental issue.

2 The Primary Film

2.1 The Inhibition of Primary Film Formation

2.1.1 Chemical Mechanisms

There is a tremendous body of literature that shows that many marine natural products exhibit antimicrobial activity in laboratory tests [60–64]. In a few of these studies the ecological role of these metabolites has been speculated upon. For most of these metabolites, the information is based on qualitative and correlative data. For example, McCaffrey and Endean [63] in their work on antimicrobial activity of tropical and subtropical sponges noted that in general, there was a negative correlation between antimicrobial activity and surface fouling. However, studies on octocorals and scleractinian corals indicate that there are elevated levels of microbial activity in the coral surface microlayer [65, 66] and that coral mucus is a nutritionally rich, easily degradable food for bacteria [67, 68].

Recent studies on natural antifouling mechanisms conducted in situ indicate that inhibition of primary film formation may not be as effective in the field as laboratory studies have suggested [65, 69]. For example, few if any macrofoulers are typically present on the gorgonian coral species *Pseudopterogorgia americana*, *Pseudopterogorgia acerosa*, and *Leptogorgia virgulata*. Unusual secondary chemistry is known from all three species [46, 50–52, 70–73]. There has been speculation from the beginning regarding the antifouling role of the secondary metabolites [46, 74]. Homarine (**1**) isolated from the gorgonian *Leptogorgia virgulata* and a furanogermacrene (**2**) from the gorgonian *Pseudopterogorgia americana* [73] have been shown to be active against benthic pennate diatoms [50, 75]. Barnacle inhibitors have also been isolated from these gorgonians [51, 52]. Despite the presence of active chemistry, in situ characterization of the gorgonian surfaces identified a bacterial film on each that ranged from 10^6 to 10^7 cells/cm^2 [76]. Time series experiments indicated that numbers of bacteria were stable over periods of several days. Experiments in different seasons also showed no change in bacterial numbers on the gorgonian surfaces [76].

1

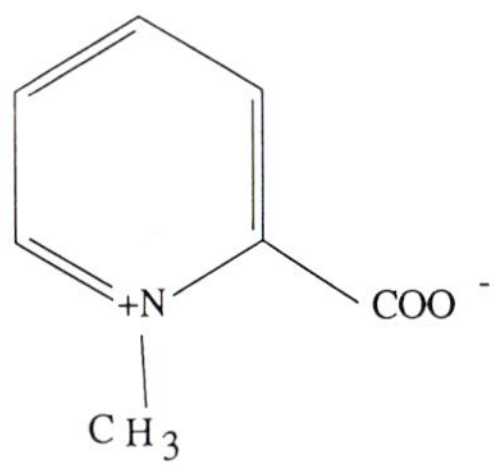

2

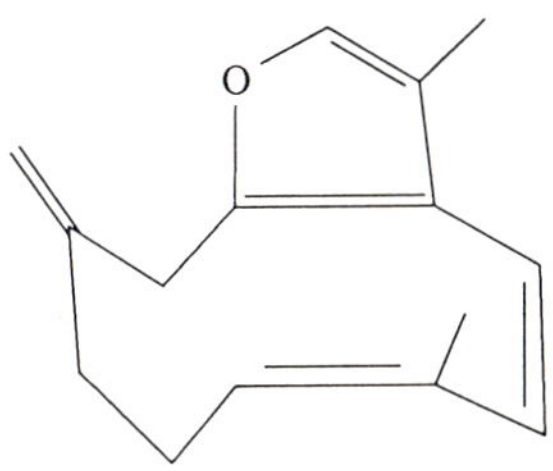

A number of studies have sought to elucidate the chemical compounds responsible for antimicrobial activity. Only those studies having done bioassays with organisms that are likely to encounter each other in the field (ecologically relevant species), are considered. Four saponins (**3**–**6**), termed muricins, were isolated from

3

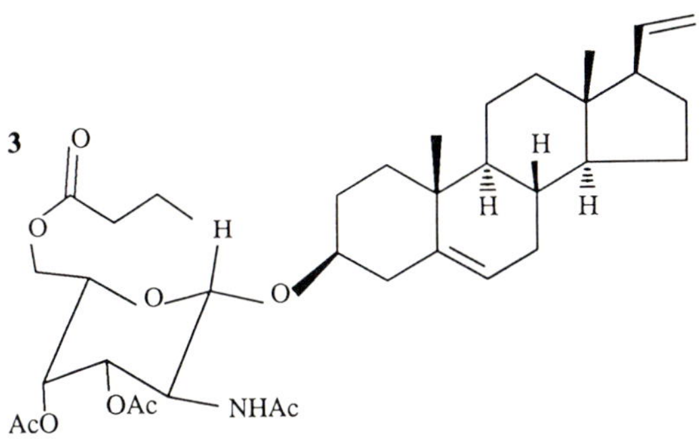

4

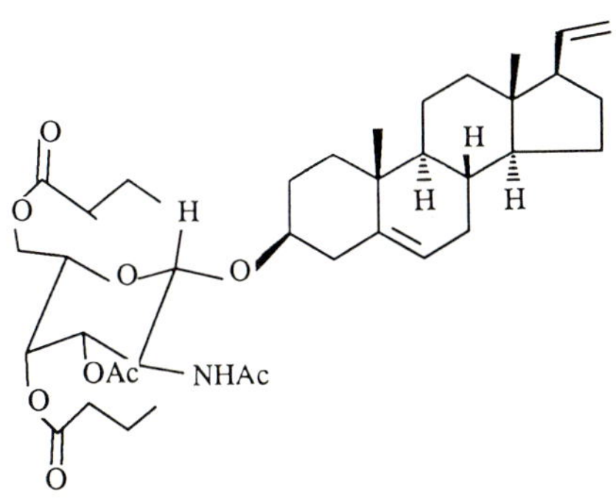

5

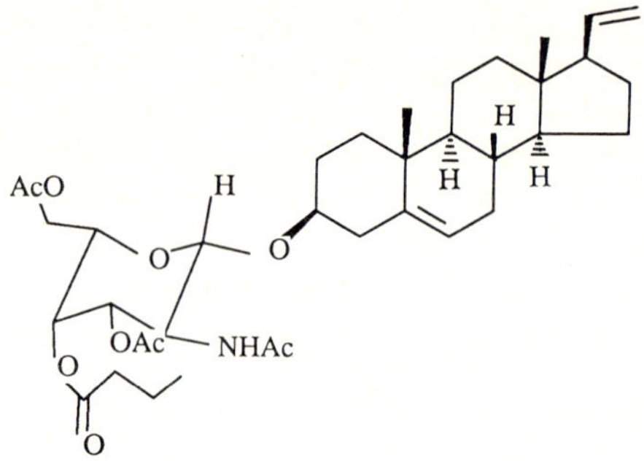

6

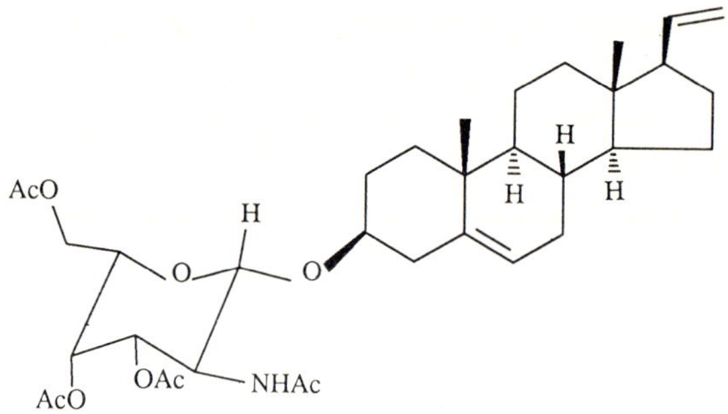

the gorgonian *Muricea californica* and proved effective in inhibiting the growth of the ditom *Phaeodactylum tricornutum* at approximately natural concentrations [61]. Yellow extracellular material (probably polyphenolics related to phloroglucinol (**7**) [77]) from the brown alga *Fucus vesiculosus* was found to be toxic to numerous marine microalgal species representing the major taxonomic groups in the marine environment [60]. The ability of water-soluble extracts from dead blades of the seagrass *Zostera marina* to inhibit growth of various pennate and centric diatoms and flagellates, has been interpreted as determining the composition and activity of organisms of living blades [78].

7

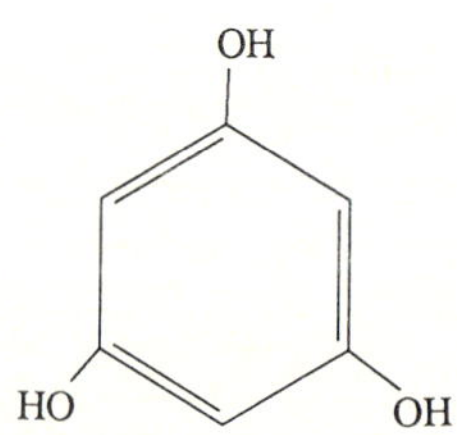

2.1.2 Other Mechanisms

Evidence for physical mechanisms which may inhibit the formation of the primary film is rather sketchy. The sloughing of surface tissues is one of the few mechanisms which has been clearly demonstrated to act upon the primary film. Diatoms and other microorganisms are shed when epidermal tissues or organic membranes are sloughed from the gorgonians *Leptogorgia virgulata* and *Pterogorgia citrina* [79], and from the brown alga *Ascophyllum nodosum* [80]. It is unclear whether these surface tissues can be ablated sufficiently quickly to inhibit the formation of the primary film, although the near epibiont-free surface of all of these organisms provides circumstantial evidence that ablation is rapid enough.

Secretion of mucus is another potential mechanism by which the formation of the primary film may be disrupted or slowed. The sloughing of mucus from surface tissues probably also serves a number of other functions including prey capture and the cleansing of sediment from organisms [81, 82]. Coelenterates show considerable variation in their production of mucus. Alcyonaceans can produce 3 to 4 orders of magnitude more mucus than some hard corals [69]. Coelenterates that release large quantities of mucus have correspondingly lower populations of bacteria associated with that mucus [67]. There exists also field evidence which suggests that mucus sheets can be shed rapidly. The soft coral *Briarium asbestinum* is believed to shed mucus sheaths at 5–10 day intervals [65]. Moreover, old mucus sheaths had significantly higher bacterial populations that newly formed sheaths, hence providing evidence of their antifouling role.

A further physical means of primary film inhibition is the surface energy of an organism. Measurements of critical surface energies for *P. americana* and *P. acerosa* [76, 83] are in the range (20–30 mN/m) that has been associated with minimal adhesiveness [84, 85]. The importance of the bacterial population on the gorgonian surface in determining the surface characteristics of these gorgonians is currently under study [86].

It is conceivable that chemical and physical mechanisms may be acting in concert; the likelihood and effectiveness of interactions between different mechanisms of antifouling defense are considered in Sect. 4.1.2.

3 Macrophytic Fouling

3.1 The Inhibition of Macrophytes

At sufficiently high levels of illumination the fouling of biological surfaces by macroalgae has been observed [34, 35]. Indeed, the great majority of fouling algae belong to macroalgal groups, and not to the Bacillariophyceae (diatoms) or Cyanophyta (blue-green algae) [34]. Despite the propensity of macroalgae to be epiphytic, there have been few attempts to examine the inhibition of fouling by macroalgae.

3.1.1 Chemical Mechanisms

A chemical comparison of epiphyte fouled and nearby unfouled colonies of the Australian alcyonarian coral, *Lobophytum pauciflorum*, in close proximity is one of the few studies of algal fouling that has included the characterization of the responsible compounds [45]. The predominant epizooic algae growing on the

8

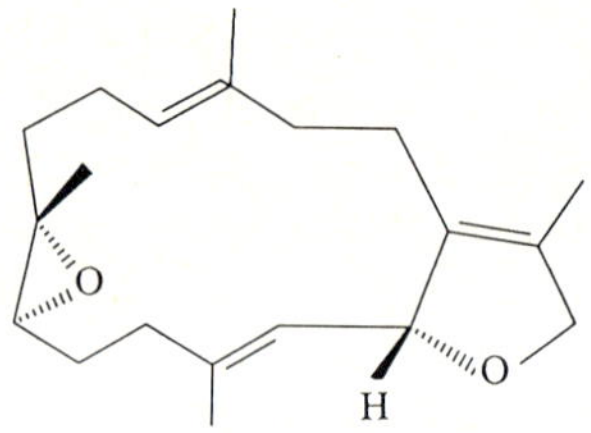

9

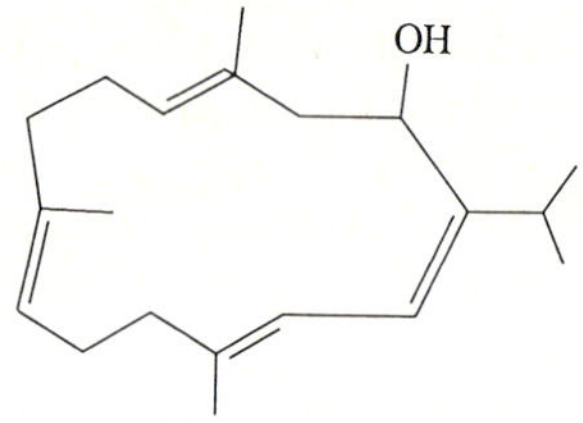

10

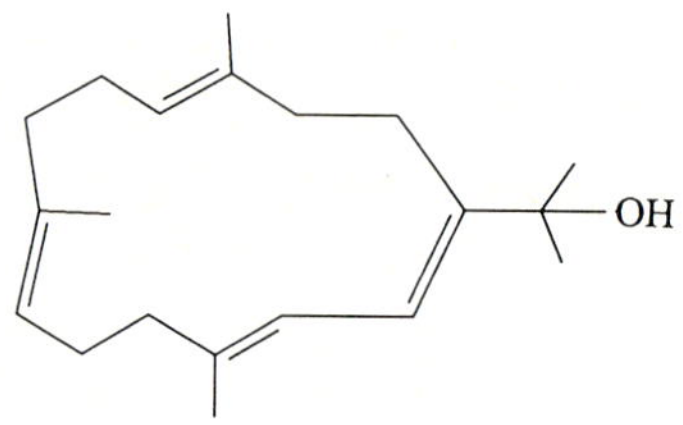

polypary of certain colonies or *L. pauciflorum* were the small filamentous algae *Ceramium flaccidum* (Kutzing) Ardissone, and *Enteromorpha* sp. The presence of these epiphytes was puzzling since soft corals generally maintain very clean polypary surfaces even under silty conditions. The overgrown colonies were found to contain the diterpene 2-*epi*-sarcophytoxide (**8**). Conspecific colonies with clean polyparies contained two structurally related diterpenes in a ratio of 3 to 1: 14-hydroxycembra-1,3,7,22-tetraene (**9**) and 15-hydroxycembra-1,3,7,11-tetraene (**10**). In assays using algal cultures of *Ceramium codii* (Richards) Mazoyer, 2-*epi*-sarcophytoxide was found to be 2–3 times more effective at inhibiting algal growth than the other two compounds. Structurally related terpenes from other soft corals were also assayed in order to understand better the structural features important for inhibition of algal growth. The presence of a five-membered dihydrofuran

11

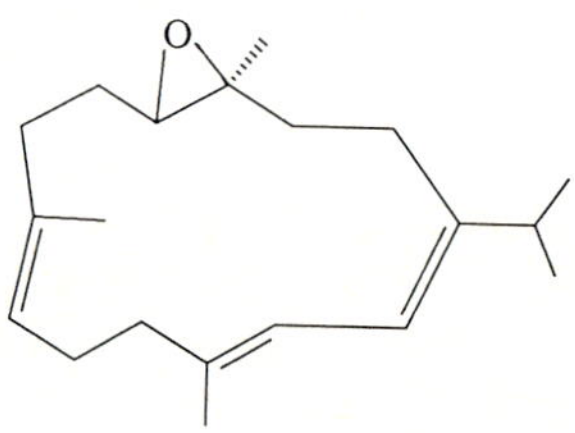

12

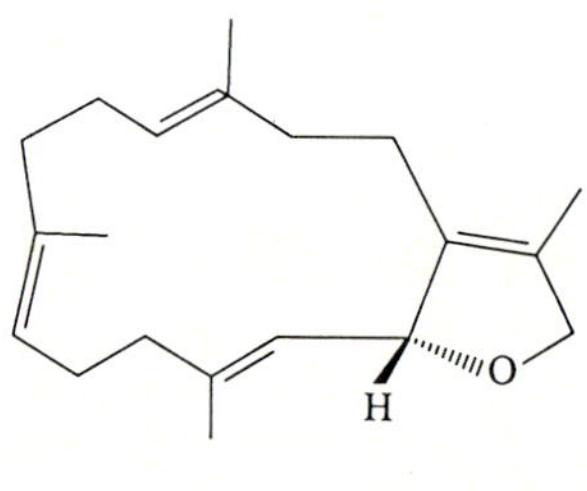

13

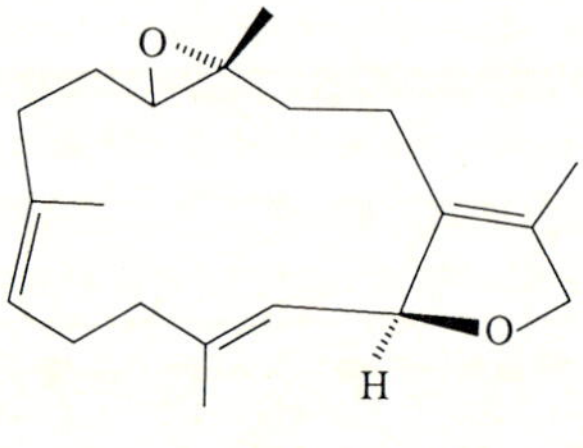

14

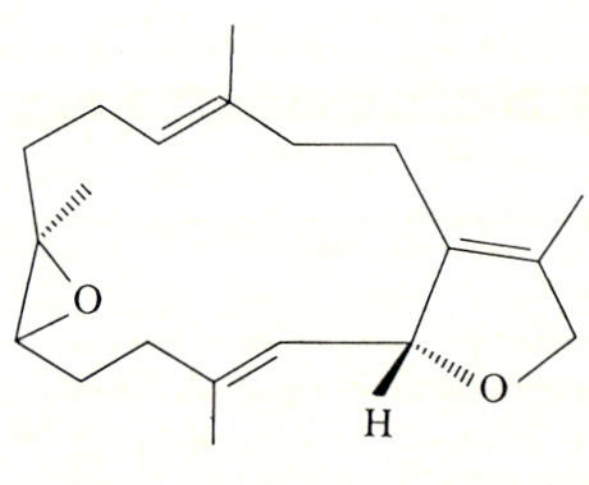

15

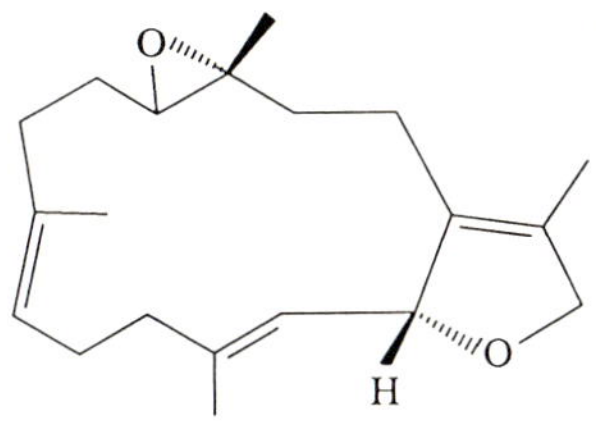

moiety and an epoxide in the same compound was more effective than compounds having only one moiety or functional group (structures **11**–**15**). The same researchers previously showed that terpenes are slowly released in situ from soft corals [48] and concluded that the terpenes described above probably play a role in the control and subsequent removal of epizooic algae from the surface of soft-coral polyparies of *L. pauciflorum.*

3.1.2 Other Mechanisms

The sloughing of surface tissues or the secretion of mucus may successfully inhibit the establishment of macrophytes. Several organisms ablate surface tissues, although the amount of tissue lost varies considerably among species. The alga *Asocphyllum nodosum* repeatedly sloughs the entire outer epidermis [80], while another fucoid, *Halidrys siliquosa,* sheds only superficial surface tissue [80]. The shedding of the outer cell walls of the alga *Enteromorpha intestinalis* may prevent macrophytes establishing [87]. Micrographs of the surface of *Chondrus crispus* convincingly demonstrate the significance of the periodic shedding of its proteinaceous cuticle in the removal of epiphytic algae [88]. We can conclude that the sloughing of surface tissues can be an effective inhibitor of macrophytes, perhaps with the exception of those epiphytic algae with deeply penetrating rhizoids (roots) [80].

Mucus may also inhibit the establishment of algal epibionts. Epiphytic algae have been reportedly trapped in mucus sheets on the surface of some alcyonacean soft corals [45].

Clearly, a physical mechanism which is effective against macrophyte establishment, such as the loss of surface tissue, is also likely to effectively inhibit the primary film and epizooites.

The removal of epiphytic macroalgae by grazing organisms [89], although unlikely to inhibit the establishment of epiphytes, may keep the densities of these plants in check.

4 Epizooic Fouling

4.1 The Inhibition of Epizooites

In many epifaunal communities, space on which to grow and reproduce is a limiting resource. The colonization of other organisms by either overgrowth or larval settlement represents a means of acquiring this resource [90–92]. Many of the nonmotile animals in these communities have developed a range of chemical and physical defenses against overgrowth and settlement by epizooites. This section will review mechanisms by which may inhibit the colonization of surfaces either by larvae or by neighboring adult organisms.

4.1.1 Chemical Mechanisms

The marine natural products reviewed in this section have shown activity against the larvae of benthic invertebrates or halted the encroachment of neighboring organisms.

4.1.1.1 Inorganic Vanadium

The ability of ascidians to accumulate vanadium has been known for some time [93] yet its biological functions remain unclear [94, 95]. Stoecker [54] noted that elevated concentrations of this metal were a metabolic poison and on observing the epibiont-free test of an ascidian *Ascidia nigra* reasoned that it might play a role in inhibiting larval settlement. Laboratory trials in which vanadium was added to agar discs demonstrated that vanadium above a concentration of 250 ppm (wet weight basis) inhibited settlement of the hydroid *Pennaria tiarella*. The length of stolons in juveniles of the ascidian *Ectinascidia turbinata* were also reduced relative to controls. Stoecker suggested that stolon growth should be a good indicator of settlement success, but offered no evidence in support of this claim.

Evidence that vanadium is an inhibitor of larval settlement is equivocal for several reasons. First, vanadium may exist in several valence states (valences 2, 3, 4, and 5) yet in most experiments the valence state of this metal is not considered [54, 96]. In the ocean, vanadium exists as V(V) [97]. It is believed that in the ascidian *Ascidia nigra*, vanadium enters vacuoles as HVO_4^{2-} [98] or $H_2VO_4^-$ [99] and is rapidly reduced to V(IV) and V(III) by tunichrome, with a final ratio of V(IV) to V(III) of 1 to 9 [97]. The mechanism of vanadium accumulation suggested by Macara and coworkers [98] accounts for the presence of H_2SO_4 in the vacuoles of *A. nigra* and is important, because V(III) and V(IV) appear to be stable only in acidic solution [94]. Elevated levels of vanadium and acid may not be a general phenomenon, however. Concentrations of H_2SO_4 up to 18% of dry mass have been recorded for the brown seaweed *Desmarestia firma* [100], yet vanadium concentrations do not appear to exceed 2.5 ppm in this alga [101].

Stoecker [54] added vanadium to agar in the form of vanadyl sulfate, hence the vanadium was initially in +4 oxidation state, but this may have changed during preparation of the agar. Furthermore, the valence state of vanadium on the surface of the animal in its natural habitat is unknown. Clearly, it is difficult to speak confidently of the significance of vanadium as an inhibitor of larval settlement without first considering the importance of oxidation state.

Second, the significance of vanadium as an inhibitor of larval settlement in ascidians is questionable since few ascidians possess vanadium in sufficiently high concentration for this metal to be an effective defense. Vanadium significantly inhibited settlement only at 250 ppm (wet weight basis), which is equivalent to more than 1000 ppm on a dry weight basis for *Ascidia nigra* [54]. Whole animal vanadium concentration for *Ascidia nigra* is a little under 2000 ppm (dry weight basis). This concentration is only exceeded by one of the 24 Bermuda ascidians for which Stoecker performed vanadium analysis: *Distaplia bermudensis* had a mean vanadium concentration of 3211 ppm (SD = 1278) [102]. Other surveys of vanadium concentration in ascidians reveal that 9% (n = 23 [103]) 22% (n = 50 [104]) of the ascidians surveyed had vanadium concentrations exceeding 2000 ppm (dry weight basis).

Ascribing a role to vanadium is further clouded by the detection of 4- to 13-fold changes in vanadium concentration during the two year interval in which the ascidian *Perophora annectens* was studied [105]. A concentration of 8000–9000 ppm was detected in the spring and summer of 1972, and 700–2000 ppm (organic dry weight) in the spring and summer of 1973. Different methods were used for these vanadium analysis but it is assumed that both methods yield comparable results. However, Swinehart and coworkers [105] also note that no variation was observed in vanadium concentration in the fluids of *Ascidia ceratodes* collected during the same time intervals. These workers do not specify the methods that were used for these determinations.

Third, to be an effective inhibitor of larval settlement, vanadium must presumably be concentrated at or near the surface of the animal. Around 63% of vanadium in *Ascidia nigra* appears to be concentrated in surface deposits [54]. This is not the case for another ascidian, *Eudistoma molle*. In this species only 20% of its total vanadium is concentrated at the surface [105]. These are the only two ascidians for which data are available on the distribution of vanadium in the surface and underlying tissues.

In contrast to her earlier contention that vanadium was an effective inhibitor of larval settlement [54] Stoecker presented evidence that the presence of epibionts was independent of vanadium concentration in ascidian tissues [102]. This discrepancy may result from limiting consideration of ascidian fouling to two groups: those with vanadium concentrations above 50 ppm and those below 50 ppm [102]. While earlier laboratory trials were only effectively inhibiting settlement at 20 times this concentration [54].

4.1.1.2 Surface Acidity

Stoecker [102] noted that the absence of epibionts among Bermuda ascidians was correlated with surface pH. She observed no epibionts on animals with an intracellular pH of 2 or less. This correlative approach was supported by laboratory trials in which she demonstrated that low pH was a potent inhibitor of the settlement of larvae and deterred fish and crabs from feeding [54, 96]. Parry [103] argued that surface acidity was not effective in deterring predators, as seawater would rapidly neutralize any acid secretions. He notes that 15 mL of seawater was sufficient to neutralize 6 crushed colonies of the acidic *Diplosoma virens*. Hence it follows that acid secretions may also be ineffective in deterring the settlement of larvae in the field. This suggestion has been borne out by field observations of Davis & Wright [106], in which they reported a heavily fouled ascidian, *Eudistoma capsulatum*, with a surface $pH \leq 2$. The solitary ascidian *Chelyosoma productum* is another good example; this species has a surface pH < 2.0, yet is heavily fouled [44]. The findings of these workers indicate that the inhibition of settlement observed by Stoecker in the laboratory was probably an artifact caused by small containers that held insufficient seawater to neutralize the acids present.

Micro and macrofouling organisms are also found on the surface of the brown algae *Desmarestia firma*, despite sulfuric acid accounting for 18% of its dry mass [100]. Unlike the situation in *E. capsulatum*, however, acid is not released from the living *D. firma* but only on the destruction of tissue [100].

4.1.1.3 Organic Compounds

Organic compounds which may inhibit the settlement of larvae have been described from a diversity of nonmotile taxa, including sponges, ascidians, bryozoans, and cnidarians. The characterization of compounds responsible for the inhibition of larval settlement ranges from testing the activity of small pieces of tissue excised from organisms to the isolation and purification of the actual compounds involved.

There is evidence that some of these organic compounds function in a nontoxic way. Young and Chia describe a water borne compound released from the ascidian *Diplosoma macdonaldi* which was not toxic to the larvae of the bryozoan *Bugula pacifica*, but significantly delayed their metamorphosis relative to controls [107]. The sponge *Xestospongia halichondroides* releases a nontoxic substance with anesthetic properties that is effective in immobilizing the larvae of a large number of invertebrate taxa [108]. A solvent extractable compound from the sea pansy *Renilla reniformis* was also nontoxic to cyprids of the barnacle *Balanus amphitrite*, but significantly inhibited their settlement [53]. This apparent nontoxic inhibition of settlement led Rittschof and his coworkers to speculate that settlement inhibitors may function by interfering with larval chemoreception, in much the same way as glucose acts in the inhibition of settlement in the larvae of the polychaete *Janua brasiliensis* [109, 110], or in the inhibition of the binding of larval lectins to bacterial polysaccharides in the case of the mollusc *Haliotis rufescens* [111]. Rittschof and coworkers have also reported that results from settlement assays with substances extracted from the gorgonian *Leptogorgia virgulata* were consistent with the inhibition of settlement via a surface-modifying mechanism [52].

Not all settlement inhibitors act in a nontoxic fashion, however. Several cembranolides (**16**–**20**) isolated from gorgonians were extremely toxic to molluscan larvae [112]. Three steroidal sulfates (**21**–**23**) have been isolated from the sponge *Toxadocia zumi* [113]. These compounds showed antimicrobial and cytotoxic activity in addition to inhibiting development of the polychaete *Salmacina tribran-*

16

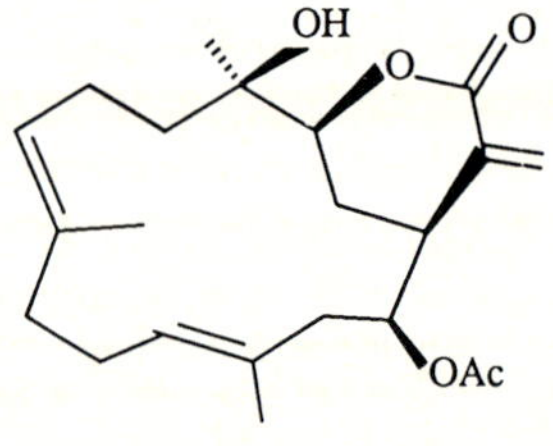

17

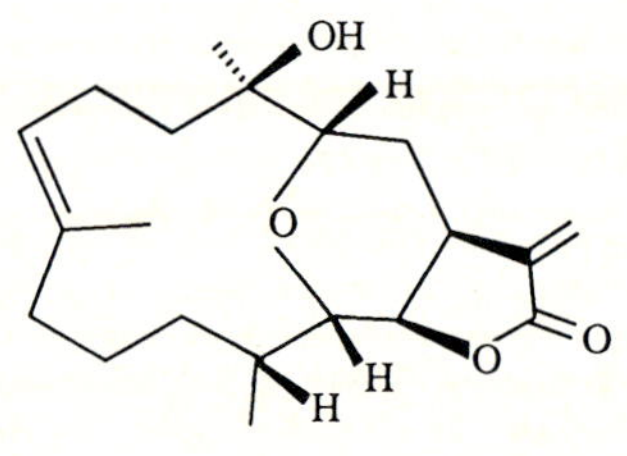

18

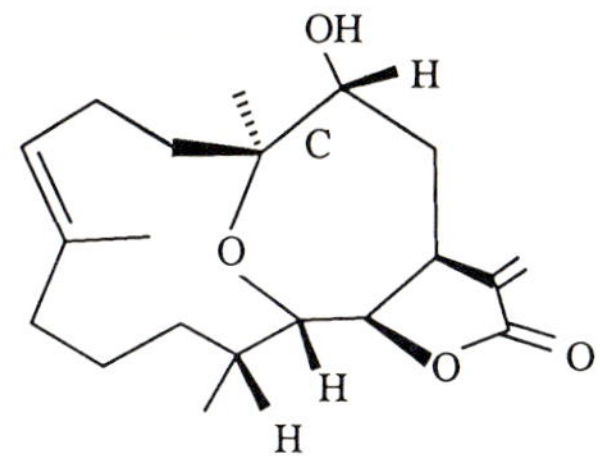

19

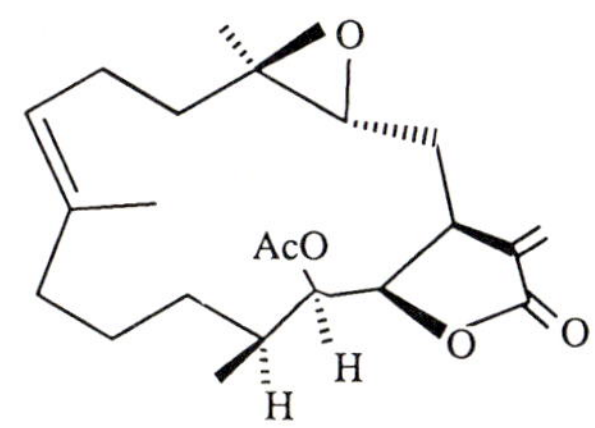

20

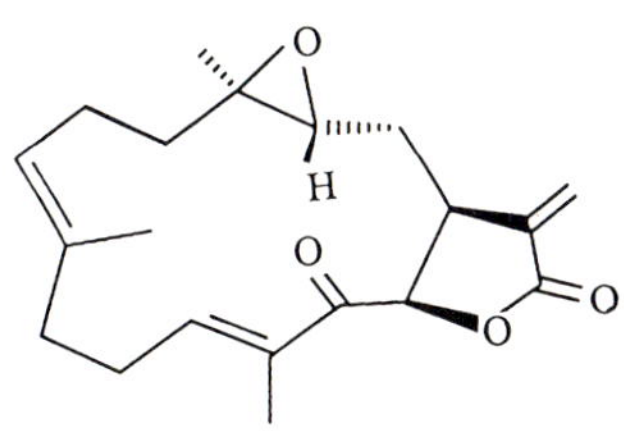

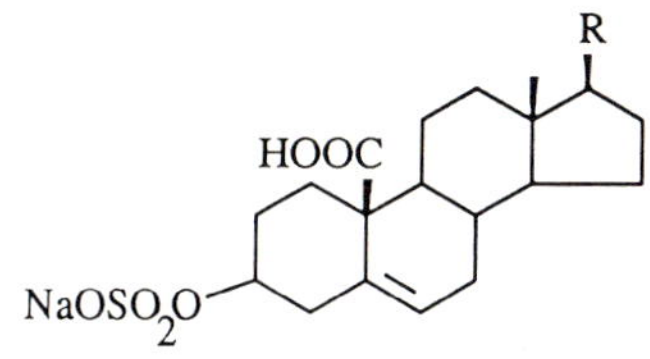

21
22
23

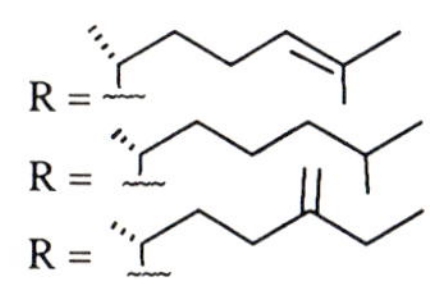

chiata. They did not affect larvae of the abalone *Haliotis rufescens* or the bryozoan *Phidolopora pacifica*. Two dibromotyrosine metabolites, aerothionin (**24**) and homoaerothionin (**25**) have been isolaed from the sponge *Aplysina fistularis* [49, 114]. In situ measurements revealed that these compounds were released into seawater and they proved to be toxic in a number of bioassays. Not only did they interfere with the feeding of a bryozoan and a hydrozoan, both potential competitors for space, but they also reduced survivorship in molluscan larvae. Notably, these compounds had no apparent effect on bryozoan and polychaete larvae. When larvae from a number of taxa have been exposed to organic compounds, differential mortality has been observed commonly [115]. For example, Dyrynda excised tissue from individuals belonging to 10 invertebrates taxa and found that tissue removed from the sponges *Cliona celata*, *Halichondria panicea*, and *Microciona atrasanguinea* and the bryozoan *Flustra foliacea* were "acutely toxic" to larvae of the bryozoan *Bugula turbinata*, yet had no effect on larvae of *Bugula flabellata* (Dyrynda defined acutely toxic as 0% settlement and 50% mortality in

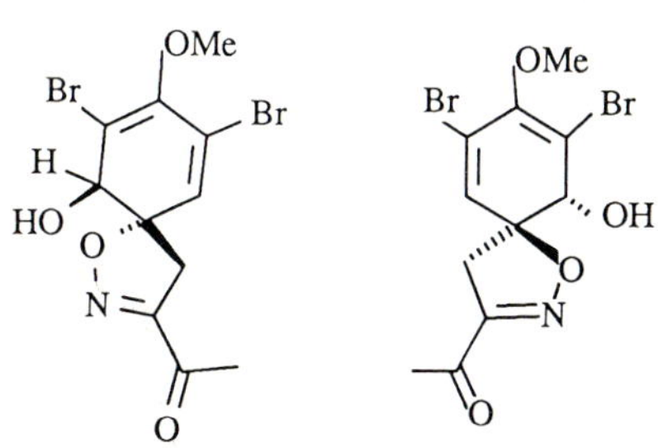

24 n = 4
25 n = 5

2 h) [116]. Rittschof and coworkers document the presence of barnacle settlement inhibitors from *Renilla reniformis* and *Leptogorgia virgulata*, which were ineffective in deterring larvae of the bryozoan, *Bugula neritina* [117]. They also found that antibryozoan settlement compounds, present in the same organisms, were ineffective against barnacle larvae. The barnacle settlement inhibitors from *Renilla reniformis* appear to be diterpene compounds, termed renillafoulins (**26**–**28**) [118].

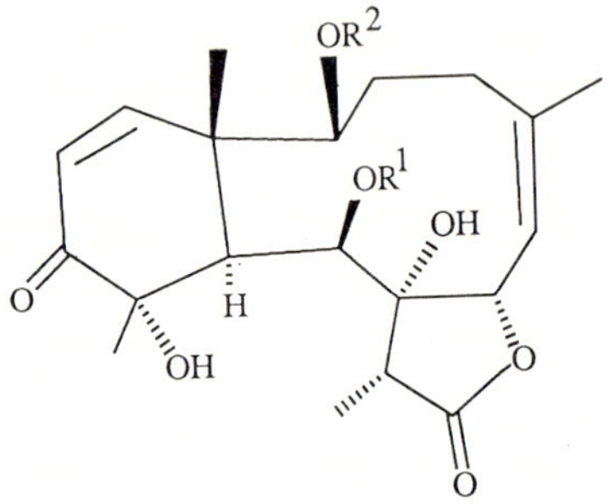

26 $R^1 = R^2 = Ac-$

27 $R^1 = AC, R^2 = C_2H_5CO-$

28 $R^1 = AC, R^2 = \underline{n}\ C_3H_7CO-$

Studies in which other larval responses to surfaces with known chemical composition were examined reveal that species respond uniquely to surfaces [115, 119–121]. Surfaces chemically altered by silanization, by adsorption of chemicals such as biologically active peptides, steroids and diterpenoids of marine origin, or by growth of a single strain of a bacterial film, generated varied but specific responses in behavior and metamorphosis assays with barnacle and bryozoan larvae [115]. Generally, chemical modifications that influenced barnacle behavior and metamorphosis did not influence bryozoan behavior and metamorphosis and vice versa. Rittschof & Costlow suggest that the ecological context of these findings is reflected in the partitioning of settlement substrata: that is, larvae from different taxa respond uniquely to surfaces [115]. This implies that surface chemistry plays a role in the distribution of adults [122].

Algae also produce toxic antifouling substances. Tannin-like substances are believed to keep the branch tips of *Sargassum natans* and *S. fluitans* free of epibionts. Extracts of young branch tips had inimical effects on common epifauna of both algal species [123].

Organisms may possess more than just one toxic or nontoxic inhibitor of larval settlement. For example, an inhibitor of settlement which was toxic to the larvae of a number of common epifauna (probably a eudistomin with a 2-pyrrolinyl substituent at C-1 on the β-carboline ring system (**29**), has been isolated from the ascidian *Eudistoma olivaceum* [106]. These workers also have evidence of a nontoxic settlement of inhibitor from the same animal [124]. Some organisms may contain settlement inducers in addition to settlement inhibitors [51, 52]. Indeed, compounds which deter larvae of some taxa may attract other larvae [112].

29

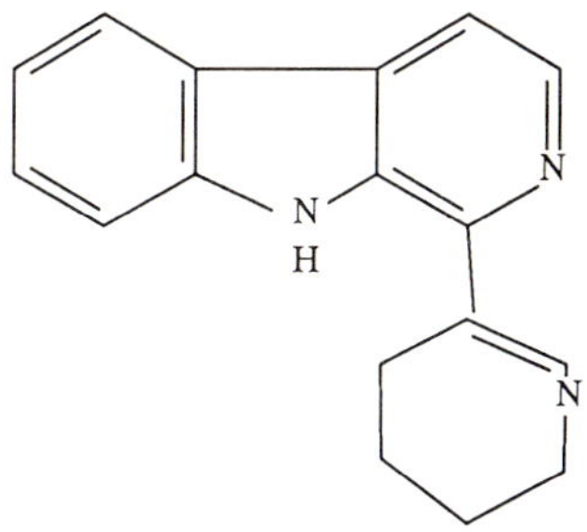

Chemical antifouling defenses are not restricted to nonmotile organisms. The hard skeletons of some mobile organisms may provide suitable sites for the settlement of invertebrate larvae. Some echinoderms contain saponins of polyketide sulfates, both of which are believed to perform an antifouling function [125, 126]. Echinoderms also have pedicillariae, which have traditionally been considered to have an antifouling function.

The presence of marine natural products at or near the surface of an organism may also play an important role in the inhibition of overgrowth by neighboring organisms and in the maintenance of living space. Zones of necrosis have been observed between the growing fronts of sponges and their competitors [90, 127–129].These field observations have been supported by field and laboratory trials in which sponges and ascidians (or their extracts) have been shown to have an inimical effect on neighboring bryozoans or corals [90, 129].

Field transplants of alcyonacean soft corals next to selected sleractinian corals resulted in high levels of local mortality in several of the hard corals [130]. One soft coral, *Lobophytum pauciflorum*, caused extensive mortality in *Porites andrewsi* in a non-contract situation. These results indicate that toxic exudates from these Great Barrier reef soft corals are responsible for the mortality observed. Indeed, the release of terpenes directly in the surrounding seawater has been reported for soft corals of the Great Barrier Reef [48, 131].

The inhibition of overgrowth is not solely restricted to chemical mechanisms; chemical and physical mechanisms often work in concert as is observed in aggressive interactions between cnidarians. Many cnidarians, particularly scleractinian corals, engage in aggressive interaction with neighboring organisms to ensure that they are not overtopped or crushed. Corals may extrude mesenteric filaments through the body wall and digest the tissues of neighboring corals [132]. The significance of this interaction to community structure and dynamics is unclear because of contradictory results obtained in different geographic areas. This response is apparently reversible in the long term [133] and indeed, aggressive coral species in the Caribbean, although present in most habitats, make up a relatively minor component of the fauna [132]. In contrast, aggressive coral species on some Indo-Pacific reefs are very abundant [134]. For example, the most aggressive coral identified on these reefs, *Galaxea clavus*, formed almost monospecific stands in some areas [134].

Cnidarians also use modified tentacles in spatial interactions [133, 135-138]. These elongated tentacles, termed sweeper tentacles in corals and catch tentacles in anemones, bear nematocysts and on contacting neighboring organisms allow the injection of venom which may prove to be fatal [139-141]. The sweeper tentacles of corals appear to be an induced response to contact with conspecifics or interspecifics [133, 142].

Sea anemones may use an additional agonistic response in the presence of neighboring organisms. Intra- and interspecific interactions among some sea anemones are mediated by the inflation of vesicular extensions at the margin of the column and their placement against neighboring organisms (acrorhagial response). In *Anthopleura elegantissima* this acrorhagial response appears to be an important mediator of competition for space [143, 144]. In *Anthopleura xanthogrammica* this reaction appears to serve as a communication mechanism between conspecifics [145]. Acrorhagial responses involve nematocysts, but the chemistry is apparently unknown [2].

For motile anemones, catch tentacles and acrorhagi appear to be important mediators of the spatial relationships between individuals. Certainly the spacing of neighboring animals is an important consideration as crowded anemones may be disadvantaged relative to uncrowded ones [27]. In view of the mobility of the organisms, these physical elaborations are probably insignificant inhibitors of overgrowth. On the other hand, sweeper tentacles may represent important inhibitors of overgrowth in nonmotile coral species.

4.1.2 Other Mechanisms

Benthic marine invertebrates may employ a wide range of physical antifouling defenses. Spines at the surface of an organism have been reported to inhibit the settlement of larvae and overgrowth by neighboring organisms. Bryozoan larvae of the genus *Bugula* may become enmeshed in spines of the sponge *Microciona atrasanguinea* which prevent their settlement [146]. Overgrowth by neighboring organisms was inhibited by the production of lateral spines in the bryozoan *Electra pilosa* [147].

Elaboration of tissue over a settled organism is a further antifouling defense. Several sponges, notably *Aplysilla sulfurea*, have been observed to engulf epibionts and it has been suggested that this may be a common form of antifouling defense in sponges, adapted to their plastic body plan and their ability to rapidly reorganize tissue [108].

Another means of physical antifouling defense is the sloughing of epidermal tissues. This mechanism was believed to be responsible for the epibiont-free test of the ascidian *Ascidia nigra*, but this assumption has not been tested [148]. The sponges *Microciona prolifera* and *Aplysilla sulfurea* allow larvae to settle on outer layers which may then be exfoliated [108]. Ablation of surface tissues or mucus has also been reported in the Cnidaria. *Leptogorgia virgulata* was observed to shed thin sheets of surface tissue in the laboratory and *Pterogorgia citrina* shed organic membranes coated with diatoms and other microorganisms in the field and in the laboratory [79]. *Briareum asbestinum* sheds mucus sheaths rapidly in the field [65]. Sloughing of surface tissue in a polycitorid ascidian has also been observed in the laboratory [149]. Fouling of algae may also be controlled by the sloughing of surface tissues [80, 86, 88].

Sloughing of epidermal tissues need not be continuous. Several ascidians have been reported to undergo cyclic phases of degeneration followed by rapid regeneration [150–152]. Degeneration represents an effective way of disposing of test material which has become fouled by microorganisms; fouling may not have been the selective pressure responsible for the evolution of degeneration, however.

Physical defenses against fouling or overgrowth need not necessarily involve the elaboration or loss of surface tissues. The extremely delicate surfaces of some organisms may render them inappropriate for larvae settlement. Some ascidians, *Diplosoma listerianum* for example, are particularly delicate [146]. The threat of overgrowth may also be reduced by the settlement of larvae, or directional growth of colonies, into habitats with reduced spatial competition [153]. The adoption of an erect growth morphology may also reduce the likelihood of overgrowth; encrusting organisms tend to grow around erect organisms rather than vertically upwards [154].

The production of mucus is a further means by which animals may inhibit the settlement of larvae or overgrowth by neighboring organisms. External mucus production is not known for bryozoans or ascidians [146], but is commonly observed for sponges and cnidarians; many produce copious quantities when they are disturbed or damaged. Mucus produced by the sponge *Myxilla* has been observed to incapacitate larvae of the bryozoan *Bugula* [146].

The production of mucus may also ensure that noxious compounds remain at or near the surface of an organism. There have been several reports of chemical defenses operating in concert with this physical antifouling defense. for example, molestation of the soft coral *Cespitularia* sp. aff. *subviridis* results in the production of mucus which contains a sesquiterpene, palustrol (**30**) [155]. This substance was found to be toxic to small crustaceans, but its effect on fouling organisms was not assessed. The gorgonian, *Pseudopterogorgia americana*, contains several heterocycles (norzooanemanin (**31**), homarine (**1**), and trigonelline (**32**) [70, 156] in addition to producing a copious slime, which comprises around 7% of the dry weight of the animal [157]. In general, sesquiterpenes are believed to function as antifouling agents [2].

30

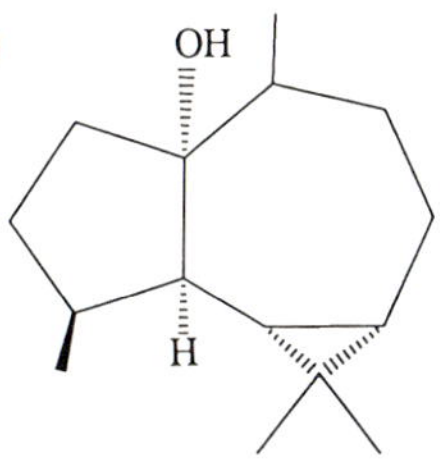

31

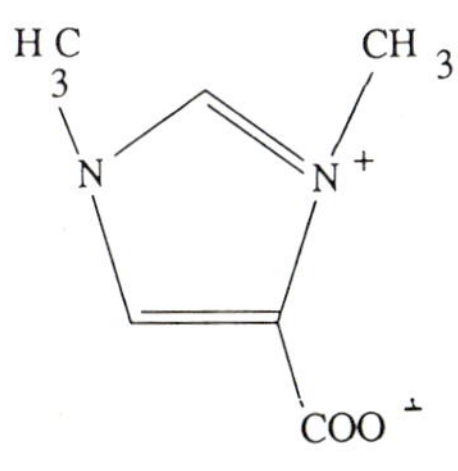

32

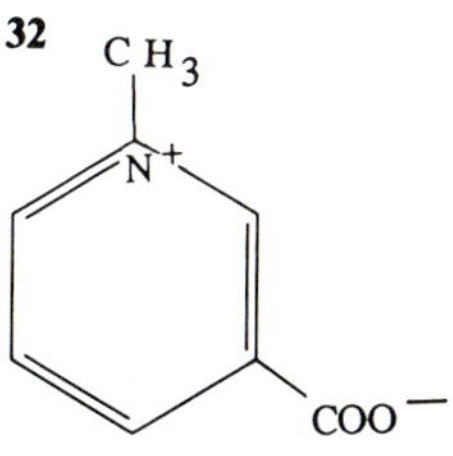

Mucus from the burrowing sponge *Siphonodictyon* sp. is believed to contain siphonodictidine (**33**), a potent inhibitor of coral growth [158]. Targett has suggested that polysaccharide mucus produced by the gorgonian *Pseudopterogorgia americana* solubilizes the nonpolar active metabolite, a furanogermacrene (**2**)

33

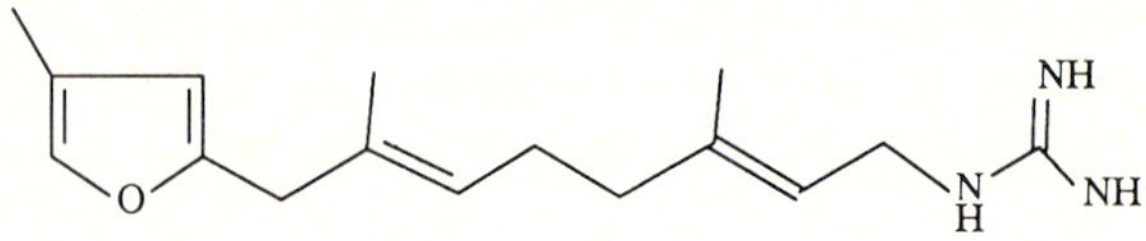

which has been shown to be responsible for inhibiting diatom growth [75]. Conversely a more polar metabolite released by the congener *P. acerosa* (which does not release mucus) may already be water-soluble [75].

The production of toxic compounds in concert with the ablation of surface tissues has been reported for two gorgonians. Homarine (**1**) has been isolated from *Leptogorgia virgulata*; it inhibits fouling by diatoms [50]. *Pterogorgia citrina*, which sheds organic membranes, may contain butenolides (**34, 35**) which have been isolated from *P. anceps* and *P. guadalupensis* [159]. Targett and coworkers have suggested that secondary compounds in *Leptogorgia* and *Pterogorgia* may slow the fouling or their surfaces until the outer surface can be sloughed off [50].

34

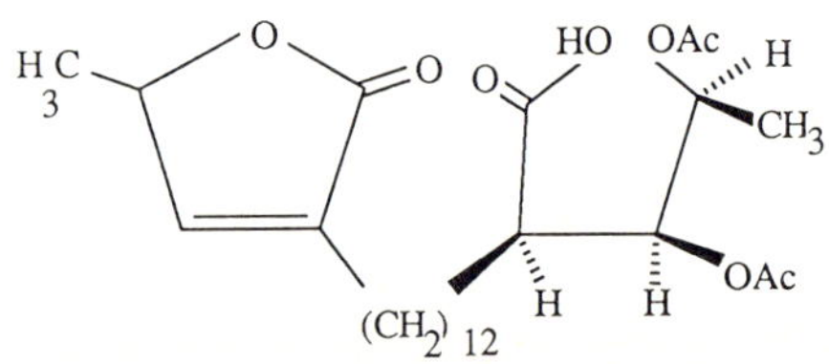

35

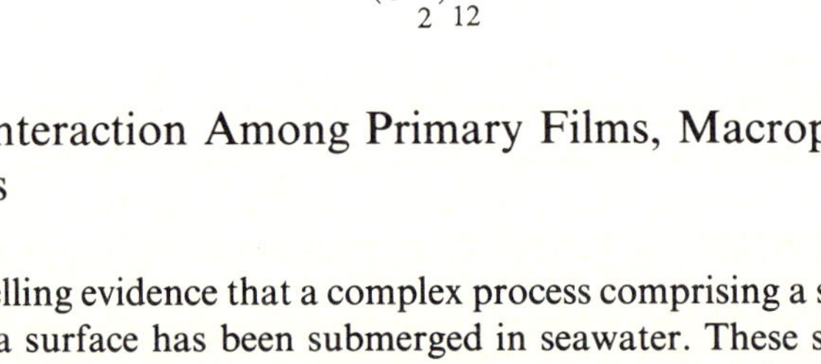

5 Fouling Interaction Among Primary Films, Macrophytes, and Epizooites

There is compelling evidence that a complex process comprising a series of stages is initiated once a surface has been submerged in seawater. These stages have been examined in detail for artificial surfaces [20, 21, 25] and for a few natural surfaces in the marine environment [76, 83, 160]. Formation of a primary film is considered an important component for subsequent biofouling on a surface, although for some macrofoulers chemical cues may mitigate its importance. Nevertheless, it appears that a variety of bacteria are always present on surfaces in the marine environment [30, 65, 69, 161] and it is increasingly evident that, in many instances, the bacterial film itself plays a role in enhancing or inhibiting further settlement. Whether the cue is positive, negative, or neutral depends upon the nature of the film, that is, its species composition, age, and the environmental conditions under which it is living; and upon the species receiving the cue [33].

Where it has been demonstrated, bacterial influence on the settlement and metamorphosis of macrofoulers has been attributed to the exopolymer which acts as an adhesive bridge in irreversibly binding bacteria to surfaces [19, 22, 162]. The polymer may be either a polysaccharide or protein [163–165], although the former is most common [166]. Studies of the hydrolyzates of marine and freshwater bacterial polysaccharides indicate that three sugars occur most frequently: D-mannose, D-glucose, and D-galactose. There is an absence of O-acetyl groups from most preparations and all appeared to contain uronic acid [163]. Because there is no single component to which an adhesive role can be assigned, Sutherland concluded that the adhesive nature of the polysaccharide was due to polysaccharide structure (e.g., 1,3 or 1,4 linkages) rather than to primary chemical composition.

For a given species of bacteria, the nature of the adhesive is dependent upon time, growth, species composition, and environmental conditions. Specific examples of bacteria and their extracellular products which influence settlement, adhesion, or metamorphosis have been reported for larvae of *Hydractinia echinata* [167], *Lytechinus pictus* [168], *Cassiopea andromeda* [169], *Bugula* spp. [170, 171], *Janua brasiliensis* [109, 172], *Crassostrea virginica* [119], and *Balanus amphitrite* [173].

Results of attachment studies of *Balanus amphitrite* larvae on surfaces to which bacteria were adsorbed varied with the bacterial species [173]. Of 18 bacterial species tested at day 1 for their effect on cyprid larvae, 7 were inhibitory, 10 had no effect and 1 was stimulatory relative to the polystyrene control. Age of the bacterial film was also a factor that influenced attachment results. Aged films of *Deleya marina* became more inhibitory to cyprid attachment, although the reasons for the increased inhibition are not known. Bacterial densities did not appear to affect cyprid attachment once they were $\geq 10^6$. These densities are comparable to those reported for sites in the natural environment [76, 174].

Larval metamorphosis of the spirorbid polychaete *Janua brasiliensis* appears to be associated with the surface of specific bacterial species [172]. The larvae settle readily on pure culture films of bacteria isolated from one of their common natural substrata, *Ulva lobata*. However, individual bacterial strains varied in their capacity to induce settlement and metamorphosis; clean slides and those coated with the diatom *Nitzchia* or with gum arabic (mainly composed of galactose), failed to induce *Janua* settlement. The active cue appears to be bound to the microbial film and is probably the polymer associated with it. Neither antibiotics nor formalin-killed films prevented settlement of competent larvae.

The rate of *Janua* metamorphosis was found to vary tremendously; however, it was determined that this was probably due to differences in quantity and quality of the polymer which in turn is a function of bacterial age and physiological status. Bacterial density was not a factor. It has been postulated that the process is mediated by lectins on the larval surface [109, 175] which recognize and bind to the glycoconjugate in the exopolymer of the bacteria. Processes mediated by lectins are blocked by low-molecular weight sugars and in experiments with *Janua* larvae, glucose, a major carbohydrate in the exopolymer of *D. marina*, significantly inhibited larval metamorphosis on surfaces coated with *D. marina*, probably by outcompeting the bacterial glycoconjugate for larval receptor sites. Experiments with periodate and trypsin suggest that the lectin glycoconjugate is a polysaccharide or glycoprotein.

The role of bacterial film in positively or negatively influencing subsequent macrofouling can be summed up as follows: 1) Bacteria can produce chemicals which act as cues that stimulate or inhibit further settlement and 2) Bacteria can modify the characteristics of a surface to enhance or inhibit further settlement. These two mechanisms are not mutually exclusive and indeed probably act in concert.

In experiments examining the role of chemistry, it is often difficult to factor out physical parameters such as texture and surface tension, because once a surface is submerged, it is modified by a host of biological, chemical, and physical processes. With *Janua*, Kirchman and coworkers [172] addressed the issue of surface characteristics and argued that, while they cannot be ignored, the data for this species suggests that surface factors are not of primary importance and that binding of larval lectins to bacterial polysaccharides is of greater significance.

More than 25 years ago pioneers of larval *ecology* demonstrated that the settlement of marine invertebrates was more rapid on surfaces bearing films of microorganisms [176]. Because the microbial film always preceeds macrofouling, many have assumed that microfouling is a necessary prerequisite to subsequent macrofouling events. Crisp maintains that many invertebrates including barnacles, tubeworms and bryozoans do prefer to settle on filmed surfaces, but will also settle on unfilmed surfaces [177]. Ascidians have also been observed to settle rapidly on unfilmed surfaces [149].

Alternatively, microfouling may be most evident initially because of the sheer numbers and rates of increase of microorganisms. Microorganisms, such as bacteria, are more common and have faster rates of increase than larger animals such as barnacles. For example, a typical bacterial population in coastal waters at 26°C has a concentration of 10^7/mL [110]. At these densities it is not surprising that bacteria are the first organisms to colonize surfaces and that the arrival of macroinvertebrates, such as barnacle cyprids, is preceeded by microorganisms. Moreover, there is often little or no evidence that later stages actually require the presence of an early stage for the successful colonization of a surface. For example, different forms of bacteria come to dominate a surface at different stages in a 'succession' [21, 25]. Small rods are usually the first colonizers [30, 161, 178, 179]; whereas stalked or filamentous forms come to dominate after a few days [30, 180]. Little has argued, "It has never been demonstrated that stalked or filamentous forms require the presence of the primary colonizing rods for their successful colonization of a surface" [24]. Moreover, stages in bacterial colonization are open to interpretation as a single bacterial species may have a variety of morphotypes, each with markedly different appearances depending on the developmental stage of the bacterium and on environmental conditions [24, 181].

The effective inhibition of the microbial film would seem unlikely to completely halt subsequent macrofouling. It follows then that organisms will require antifouling defenses which are effective against a broad taxonomic range of micro- and macrofouling organisms, and it is not surprising that many organisms possess a combination of chemical and physical antifouling defenses.

Interactions between macrophytes and epizooites during the fouling process have not, as far as we are aware, been examined. The few studies that have examined interactions between algae and invertebrates on natural rock surfaces indicate that the outcomes of these interactions are highly complex and may be dependent on environmentally induced physical stress, water movement at the study site, and biotic interactions with other grazing species [151, 182–184].

6 Ecological and Evolutionary Considerations

6.1 The Advantages and Disadvantages of Being Fouled

An organism will clearly be disadvantaged if the presence of epibionts impedes its ability to capture resources such as light or food. Feeding efficiency may be reduced by obstructing or occluding feeding structures; alternately, epibionts may compete directly with the host organism for food. Yet surprising, complete overgrowth need not result in the death of the overgrown organism [185], although a heavy load of fouling organisms may increase the chances of the host organism being dislodged by water movements. Under any of these circumstances selection pressure should be strong for the development of an antifouling defense. Presumably the production of that defense will incur a cost of the organism. Inducing the bryozoan, *Membranipora membranacea*, to produce spines as an antipredator defense resulted in a measurable reduction in colony growth rate which translated to a decrease in sexually produced propagules [186]. Chemical defense can also be metabolically costly, although these costs can be reduced and even eliminated if the production of the chemical defense is the byproduct of a primary metabolic pathway. It would be maladaptive for a defense to be maintained if the benefits did not outweigh the costs.

The presence of epibionts need not always be disadvantageous. A heavy coating of fouling organisms may effectively camouflage an organism [187], although any benefits accruing would be balanced against the increased likelihood of being dislodged by water movement and decreased resources.

6.2 The Misinterpretation of data

Because the settlement preferences of larvae in situ often cannot be observed, inferences about settlement preferences must be drawn from the distribution of adults. Reliance on inference is sometimes unavoidable, but it may result in misinterpretation. Patterns exhibited by adults and even juveniles of a species may reflect differential settlement, differential survival or a combination of the two, as Davis demonstrated with direct field observations [40, and see 42]. For example the polychaete *Janua brasiliensis* is more common on the older tips of the seagrass

Zostera marina than the younger bases [188]. Based on this distribution of the adult worms Nelson made inferences about the settlement preferences of these polychaetes and ignored the potential role of differential mortality in creating the patterns of distribution that he observed. Unfortunately such oversights are often promulgated and in reinterpreting Nelson's observations. Mitchell and Kirchman suggested that it was bacteria and not diatoms which induced settlement and produced the observed patterns of adult distribution [110]. The alternate hypothesis, that differential mortality may also play a role in determining the distribution of *J. brasiliensis* on seagrass blades, was never rigorously tested.

Potential also exists for incorrectly ascribing settlement of larvae to a particular site as larval preference, when it may in fact be larval avoidance or vice versa. For example, Al-Ogily and Knight-Jones noted that larval settlement was negatively correlated with antibiotic activity for the frondose bryozoan, *Flustra foliacea* and the brown alga *Laminaria* sp. [189]. The implication was that larvae were avoiding the antibiotic activity. Causality cannot be implied from a correlation and care must be taken to examine alternate explanations. In this case, there are other clear advantages for larvae which settle preferentially on the youngest sections of a substratum; for example, spatial competition is likely to be less intense [190, 191]. Moreover, other settlement cues, which have nothing to do with the antibiotic activity, may cause larvae to show preference for a particular site on a frond or avoidance of another site. Before the importance of any one cue is assigned, others likely to be of importance must also be tested.

Caution must also be exercised in proposing antifouling roles for compounds isolated from marine invertebrates, unless there is evidence that the compound is released from, or concentrated at, the surface of the organism. Moreover, interpretations based wholly on laboratory assays may suffer from artifacts, such as static water and small container size.

6.3 Variation in Antifouling Defenses

In the absence of variation, ecological relationships could not evolve. Variation in the possession and allocation of plant chemical defenses is commonplace in terrestrial ecosystems [192] and recent work with marine algae indicates that variation within and between individuals and across habitats may also commonly occur [193–195]. In contrast, there is a general paucity of quantitative data available on variation in antifouling defenses; most references to variation are purely anecdotal. The presence of active metabolites from the sponge *Aplysina fistularis* was found to be influenced by depth [49]. Aerothionin and homoaerothionin were isolated from specimens collected at 0.5 m at Casa Cove, La Jolla, California, USA, but samples from between 5- and 15-m depths at this site lacked these metabolites [62].

Differences between sites have also been recorded. The colonial ascidian *Eudistoma olivaceum* contains up to 20 eudistomins [196, 197]. At least one of these alkaloids possesses potent antifouling properties [124]. Of the 20 eudistomins which have been isolated from this ascidian, it appears that the actual mixture of eudistomins present is dependent on the geographic area from which specimens were collected; For example, eudistomins K and L were isolated from *Eudistoma olivaceum* collected in Florida, but not from specimens collected in Belize [196].

Overlying the spatial variation which may be present is the likelihood of temporal variation. For example, antibiosis in sponges near Cancun, Mexico, appears to have a seasonal component [2].

7 Conclusions and Directions for Future Research

The sea has proven to be a rich source of natural products; over 1400 have been isolated from marine invertebrates and algae within the last decade [14–16] and many of these are biologically active [11, 198]. The ecological role (if any) of many of these compounds is completely unknown. Clearly, the primary objective of chemical ecology is to understand the significance of marine natural products in mediating the interactions between organisms. If chemical ecology is in its infancy, then the role of marine natural products in inhibiting the epibiosis of organisms may be considered newborn. In this young but rapidly expanding field, several important directions for future research can be identified.

Marine ecosystems are usually diverse; defenses which are effective against one organism may not be effective against another. Indeed, it is appropriate to consider if one compound can fulfill more than one defensive role, say as an antifeedant as well as an antifouling compound. Clearly, in order to correctly interpret the role of natural products in mediating interspecific interactions, we strongly recommend that bioassays utilize a range of ecologically relevant species.

There is good evidence that exogenous biochemical cues or triggers can regulate the settlement and metamorphosis of marine invertebrates [111, 199, 200]. Yet for antifouling compounds, we are aware of only one study which has examined which portion of the compound exhibits the biological activity [45]. This avenue has been fruitfully pursued in the examination of facilitation of larval settlement [111, 201].

In marine ecosystems, observations of temporal and spatial variation in the possession of chemical antifouling defenses within a species have been largely casual. Detailed quantification of spatial and temporal variation in chemical antifouling defenses, and how these factors might influence interactions between organisms should have a high priority for research.

The possession of toxic antifouling defenses by several organisms poses a number of questions. For example, where are these compounds stored? How are they stored and disseminated without damaging the host organism? What controls the production and release of these compounds? How costly are they in terms of energetics?

Inducible antipredator defenses have been demonstrated repeatedly in terrestrial ecosystems and are now being documented in temperate and tropical marine habitats [202, 203]. We are unaware of any current studies demonstrating inducible antifouling defenses. This area certainly warrants further investigation.

Throughout this review we have considered the consequences of epibiosis to the host organism. It would clearly also be of interest to examine the consequences of such interactions to the epibionts. The scant date which are available indicate that epibiosis may indeed represent a spatial refuge from predators [44], competitors, or physical sources of mortality.

In shaping the patterns of distribution and abundance of individuals, chemical ecology has important implications for the structure and dynamics of algal and invertebrate communities [116, 184, 204]. The fouling of living surfaces represents a means of invading a community in which spatial resources are extremely limited. Hence chemical and physical antifouling defenses may determine which epibionts may invade a community and when they may do so.

In their recent overview of chemical ecology, Bakus & coauthors argued that "natural products chemistry has little meaning without a framework based on ecological and evolutionary theory" [2]. We concur with these authors and believe that collaboration between chemists and ecologists in an interdisciplinary approach to chemical ecology will prove to be most fruitful. Perhaps the most valuable contribution the chemists and ecologists can make to this field is a reduction in the parochiality of thought. As Stephen Jay Gould insightfully observed "We all have a tendency to spin universal theories from a limited domain of surrounding circumstances. Many geneticists read the entire world of evolution in the confines of a laboratory bottle filled with fruit flies" [205]. It is inappropriate to consider chemical mechanisms of antifouling defense in isolation, as many organisms possess physical antifouling defenses, many which may act in concert with chemical defenses. Chemical compounds may also interact synergistically [196]. Indeed, the true effectiveness of antifouling defenses may only be determined when the overall antifouling "strategy" of an organism has been assembled [75, 76, 83]. The presence of antifouling defenses across a broad range of non-motile taxa argues strongly for biofouling as potent selective pressure molding the defenses of these plants and animals.

Acknowledgments. The authors thank Rod Fujita, Brian Bingham, and Nick Vrolijk for their comments on early drafts of this chapter. Barbara Sampson, Doreen Keller, and Jane Davis kindly assisted with the production of the figures and the typescript. Financial support was provided by a Harbor Branch Institution Postdoctoral Fellowship to ARD and an ONR Research Contract (N00014-88-K-0290) to NMT. This is contribution number 676 from the Harbor Branch Oceanographic Institution.

8 Glossary

Cyprid	The larval stage of a barnacle which is capable of settling and undergoing metamorphosis
Exopolymer	Extracellular polymer produced by an organism, often a polysaccharide. These vary with species, physiological condition and growth condition

Lectins	Proteins that bind to specific carbohydrate moieties
Glycoconjugate	Chemical component of the exopolymer to which the lectin binds
Polypary	The supporting structure to which each member of a colony of polyps is attached
Silanization	Process by which organo-functional silanes bond organic polymer systems to inorganic substrates
Surface energy	An estimate of substratum wettability
Test	Thick leathery exterior of ascidians

References

1. Whittaker RH, Feeny PP (1971) Science 171:757
2. Bakus GJ, Targett NM, Schulte B (1986) J. Chem. Ecol. 12:951
3. Rosenthal GA, Janzen DH (1979) Herbivores: their interaction with secondary plant metabolites. Academic Press, New York
4. Norris JN, Fenical W (1982) in: Rutzler K. MacIntyre IO (eds) The Atlantic Barrier Reef Ecosystem at Carrie Boy Cay, Belize 1: Structure and Communities. Smithsonian Contributions to the Marine Sciences. 539 pp.
5. Coley PD, Bryant JP, Chapin III FS (1985) Science 230:895
6. Paul VJ, Hay ME (1986) Mar. Ecol. Prog. Ser. 33:255
7. Bazzaz FA, Chiariello NR, Coley PD, Pitelka LF (1987) Bioscience 37:58
8. Hay ME, Fenical W (1988) Ann. Rev. Ecol. Syst. 19:111
9. Kittredge JS, Takahashi FT, Lindsey J, Lasker R (1974) Fish. Bull. US 74:1
10. Herring PJ (1979) Pure Appl. Chem. 51:1901
11. Scheuer PJ (ed) (1978–1983) Marine natural products: chemical and biological perspectives, Vols 1–5, New York, Academic Press
12. Fenical W (1982) Science 215:923
13. Faulkner DJ (1984a) Nat. Prod. Rep. 1:251
14. Faulkner DJ (1984b) Nat. Prod. Rep. 1:551
15. Faulkner DJ (1986) Nat. Prod. Rep. 3:1
16. Faulkner DJ (1987) Nat. Prod. Rep. 4:539
17. Christophersen C, Anthoni U (1986) Sulphur Rep. 4:365
18. Marshall, KC (1980) in: Berkeley RCW, Lynch JM, Melling J, Rutter PR, Vincent B (eds) Microbial Adhesion to Surfaces. Ellis Horwood Ltd, Chichester
19. Characklis WG, Cooksey KE (1983) Adv. Appl. Microbiol. 29:93
20. Baier RE (1984) in: Costlow JD, Tipper RC (eds) Marine biodeterioration: an interdisciplinary study, Naval Institute Press, Annapolis, Maryland, pp 57–62
21. Savage DC, Fletcher M (1985) Bacterial Adhesion: Mechanism and Physiological Significance, Plenum Press, New York
22. Hamilton WA (1987) Biofilms: microbial interactions and metabolic activities. SGM 41, Ecology of Microbial Communities. Cambridge University Press
23. Costlow JD, Tipper RC (1984) (eds) Marine biodeterioration: an interdisciplinary study. Naval Institute Press, Annapolis, Maryland
24. Little BJ (1984) in: Costlow JD, Tipper RC (eds) Marine biodeterioration: an interdisciplinary study, Naval Institute Press, Annapolis, Maryland, pp 63–67

25. Thompson MFD, Sarojini R, Nagabhushanam R (1988) Marine biodeterioration. Oxford and IBH Publishing Co., New Delhi
26. Connell JH, Keough MJ (1985) in: STA Pickett, PS White (eds) The ecology of natural disturbance and patch dynamics. Academic Press, New York, USA, pp 125–151
27. Sebens KP (1983) J. Exp. Mar. Biol. Ecol. 71:73
28. Gerrodette T (1981) Ecology 62:611
29. Kay AM, Keough MJ (1981) Oecologia 48:123
30. Dempsey, MJ (1981) Mar. Biol. 61:305
31. Paerl H (1985) in: Savage DC, Fletcher M (eds) Bacterial Adhesion. Plenum Publishing Co., New York
32. Hauser D, Levandowsky M, Hutner S, Churosoff L, Hollwitz J (1975) Microb. Ecol. 1:246
33. Marshall KC (1985) in: Savage DC, Fletcher M (eds) Bacterial Adhesion. Plenum Publishing Co., New York
34. Fletcher RL, Jones AM, Jones EBG (1984) in: Costlow JD, Tipper RC (eds) Marine biodeterioration: an interdisciplinary study. Naval Institute Press, Annapolis, Maryland, pp 172-182
35. Burke RD (1983) Can. J. Zool. 61:1701
36. Meadows PS, Campbell JI (1972) Adv. Mar. Biol. 10:271
37. Scheltema RS (1974) Thalassia Jugosl. 10:263
38. Moore PG (1975) Mar. Behav. Physiol. 3:97
39. Crisp DJ (1976) in: Newell RC (ed) Adaptations to environment: essays on the physiology of marine animals. Butterworths, London, pp 83–124
40. Davis AR (1987) J. Exp. Mar. Biol. Ecol. 106:57
41. Butman CA (1987) Oceanogr. Mar. Biol. Ann. Rev. 25:113
42. McGee BL, Targett NM (1989) J. Exp. Mar. Biol. Ecol. in press
43. Keough MJ (1984) Ecology 65:677
44. Young CM (1986) Mar. Biol. 91:513
45. Coll JC, Price I, Konig GM, Bowden BF (1987) Mar. Biol. 96:129
46. Burkholder PR (1973) in: Jones OA, Endean R (eds) Biology and geology of coral reefs. Vol. 2. Academic Press, New York, pp 117–203
47. Young CM, Chia FS (1987) in: Giese AC, Pearse JS, Pearse VB (eds) Reproduction of marine invertebrates. Vol. 9. General aspects: seeking unity in diversity. Blackwell/Boxwood, Palo Alto and Pacific Grove, California, pp 385–463
48. Coll JC, Bowden BR, Tapiolas DM (1982) J. Exp. Mar. Biol. Ecol. 60:293
49. Thompson JE (1985) Mar. Biol. 88:23
50. Targett NM, Bishop SS, McConnell OJ, Yoder JA (1983) J. Chem. Ecol. 9:817
51. Standing J, Hooper IR, Costlow JD (1984) J. Chem. Ecol. 10:823
52. Rittschof D, Hooper IR, Branscomb ES, Costlow JD (1985) J. Chem. Ecol. 11:551
53. Rittschof D, Hooper IR, Costlow JD (1986) Bull. Mar. Sci. 39:376
54. Stoecker D (1978) Biol. Bull. 155:615
55. Alzieu C, Thibaud Y, Heral M, Boutier B (1981) Rev. Trav. Inst. Peches. Marit. 44:301
56. Alzieu C, Sanjuan J, Deltreil JP, Borel M (1986) Mar. Pollut. Bull. 17:494
57. Waldock MJ, Thain JE (1983) Mar. Pollut. Bull. 14:411
58. Bryan GW, Gibbs PE, Hummerstone LG, Burt OR (1986) J. Mar. Biol. Assoc. UK 66:611
59. Laughlin RB, Gustafson R, Pendoley P (1988) Mar. Ecol. Prog. Ser. 48:29
60. McLachian J, Craigie JS (1964) Can. J. Bot. 42:287
61. Bandurraga MM, Fenical W (1985) Tetrahedron 41:1057

62. Thompson JE, Walker RP, Faulkner DJ (1985) Mar. Biol. 88:11
63. McCaffrey EJ, Endean R (1985) Mar. Biol. 89:1
64. Amade P, Charroin C, Baby C, Vacelet J (1987) Mar. Biol. 94:271
65. Rublee PA, Lasker HR, Gottfried M, Roman MR (1980) Bull. Mar. Sci. 30:888
66. Paul JH, DeFlaun MF, Jeffrey WH (1986) Mar. Ecol. Prog. Ser. 33:29
67. Ducklow HW, Mitchell R (1976b) Limnol. Oceanogr. 24:715
68. Herndl BJ, Velimirov B (1986) Mar. Biol. 90:363
69. Ducklow HW, Mitchell R (1979a) Limnol. Oceanogr. 24:706
70. Weinheimer AJ, Washecheck PH, Van Der Helm D, Bilayet-Hossain M (1968) Chem. Commun. (18):1070
71. Ciereszko LS, Karns TKB (1973) in: Jones OA, Endean R (eds) Biology and geology of coral reefs. Vol 2. Academic Press, New York, p 183
72. Bandurraga MM, McKittrick B, Fenical W, Arnold E, Clardy J (1982) Tetrahedron 38:305
73. Izac RB, Bandurraga MM, Wasylyk JM, Dunn FW, Fenical W (1982) Tetrahedron 38:301
74. Ciereszko LS, Sifford DH, Weinheimer AJ (1960) Ann. N. Y. Acad. Sci. 90:917
75. Targett NM (1988) in: Thompson MFD, Sarojini R, Nagabhushanam R (eds) Marine biodeterioration. Oxford and IBH Publishing Co., New Delhi, pp 609–617
76. Vrolijk NH (1988) A Model of Synergistic Natural Antifouling Defenses in Gorgonian Corals. M.S. Thesis, University of Delaware. 92pp.
77. Ragan MA, Glombitza K-W (1986) in: Round FE, Chapman DJ (eds) Progress in Phycological Research, Vol 4, Biopress Ltd, Bristol, England, pp 129–241
78. Harrison PG, Chan AT (1980) Mar. Biol. 61:21
79. Patton WK (1972) Bull. Mar. Sci. 22:419
80. Filion-Myklebust C, Norton TA (1981) Mar. Biol. Lett. 2:45
81. Hubbard JA, Pocock YP (1972) Geol. Rundsch. 61:598
82. Lewis JB, Price WS (1976) J. Zool. Lond. 178:77
83. Vrolijk NH, Targett NM (1989) Biofouling in press
84. Baier RE (1972) in: Acker RF, Brown BF, DePalma JR, Iverson WP (eds) Proceedings of the 3rd International congress on marine corrosion and fouling, Evanston, IL, Northwestern University Press, pp. 633–639
85. Baier RE (1980) in: Bitton G, Marshall KC (eds) Adsorption of microorganisms to surfaces, New York, John Wiley and Sons, pp 59–104
86. Targett NM, work in progress
87. McArthur DM, Moss BL (1977) Br. Phycol. 12:359
88. Sieburth JM (1975) Microbial Seascapes, University Park Press, Baltimore
89. Potts DC (1977) J. Exp. Mar. Biol. Ecol. 28:207
90. Jackson JBC, Buss L (1975) Proc. Nat. Acad. Sci. 72:5160
91. Sutherland JP, Karlson RH (1977) Ecol. Monogr. 47:425
92. Sutherland JP (1981) Am. Nat. 118:499
93. Henze M (1911) Physiol. Chem. 72:494
94. Macara IG (1980) Trends in Biochem. Sci. 5:92
95. Rowley AF (1983) J. Exp. Zool. 227:319
96. Stoecker D (1980b) Ecology 61:1327
97. Bruening RC, Olitz EM, Furukawa J, Nakanishi K, Kustin K (1986) J. Nat. Prod. 49:193
98. Macara IG, McLeod GC, Kustin K (1979) Biochem. J. 181:457
99. Dingley AL, Kustin K, Macara IG, McLeod GC (1981) Biochim. Biophys. Acta 649:493

100. Anderson RJ, Velimirov B (1982) PSNZI: Mar. Ecol.3:357
101. Cockerill BM, Finch P, Percival E (1978) Phytochem. 17:2129
102. Stoecker D (1980a) Mar. Ecol. Prog. Ser. 3:257
103. Parry DL (1984) Mar. Ecol. Prog. Ser. 17:279
104. Hawkins CJ, Kott P, Parry DL, Swinehart JH (1983) Comp. Biochem. Physiol. 76:555
105. Swinehart JH, Biggs WR, Halko DJ, Schroeder NC (1974) Biol. Bull. 146:302
106. Davis AR, Wright AE (1988) Mar. Biol., in press
107. Young CM, Chia FS (1981) Int. J. Invert. Reprod. 3:221
108. Ware GN (1984) The patterns and mechanisms of antifouling in some temperate marine sponges. Doctoral Thesis, Duke University
109. Kirchman D, Graham S, Reish D, Mitchell R (1982b) Mar. Biol. Lett. 3:131
110. Mitchell R, Kirchman D (1984) in: Costlow JD, Tipper RC (eds) Marine biodeterioration: an interdisciplinary study. Naval Institute Press, Annapolis, Maryland, pp 49–56
111. Morse D (1984) in: Costlow JD, TIpper RC (eds) Marine biodeterioration: an interdisciplinary study. Naval Institute Press, Annapolis, Maryland, pp 134–140
112. Hadfield MG, Ciereszko LS (1978) in: Kaul PN, Sindermann CJ (eds) Drugs and food from the sea: myth or reality? University of Oklahoma Press, pp 145–150
113. Nakatsu T, Walker RP, Thompson JE, Faulkner DJ (1983) Experientia 39:759
114. Walker RP, Thompson JE, Faulkner DJ (1985) Mar. Biol. 88:27
115. Rittschof D, Costlow JD (in press a) in: Proceedings of the 21st European Marine Biology Symposium
116. Dyrynda PEJ (1985) in: Gibbs PE (ed) Proceedings of the 19th European Marine Biology Symposium. Cambridge University Press, Cambridge, pp 411–421
117. Rittschof D, Hooper IR, Costlow JD (1988) in: Thompson MF, Sarojini R, Nagabhushanam R (eds) Marine biodeterioration. Oxford and IBJ Publishing Co., New Delhi, pp 599–608
118. Keifer PA, Rinehart KL, Hooper IR (1986) J. Org. Chem. 51:4450
119. Weiner RM, Segall AM, Colwell RR (1985) Appl. Envir. Microbiol. 49:83
120. Coon SL, Bonar DB and Weiner RM (1985) J. Exp. Mar. Biol. Ecol. 94:211
121. Rittschof D, Costlow JD (in press b) in: Proceedings of the 22nd European Marine Biology Symposium
122. Strathmann RR, Branscomb ES, Vedder K (1981) Oecologia 48:13
123. Sieburth JM, Conover JT (1965) Nature 208:52
124. Davis AR, Wright AE (1988) J. Chem. Ecol. in press
125. Rideout JA, Smith NB, Sutherland MD (1979) Experientia 35:1273
126. Russell FE (1984) Adv. Mar. Biol. 21:59
127. Bryan PG (1973) Micronesica 9:237
128. Ayling AL (1983) Biol. Bull. 165:343
129. Porter JW, Targett NM (1988) Biol. Bull. 175:230
130. Sammarco PJ, Coll JC, La Barre S, Willis B (1982) Coral Reefs 1:173
131. Coll JC, Bowden BF, Tapiolas DM, Willis RH, Dyura P, Streamer M, Trott L, Webb L (1985) Tetrahedron 41:1085
132. Lang J (1973) Bull. Mar. Sci. 23:260
133. Wellington GM (1980) Oecologia 47:340
134. Sheppard CRC (1979) Mar. Ecol. Prog. Serv. 1:237
135. Williams RB (1975) J. Nat. Hist. 9:241–248
136. den Hartog JC (1977) in: Proceedings of the 3rd International Coral Reef Symposium, University of Miami, Florida, pp 463–469
137. Purcell JE (1977) Biol. Bull. 153:335

138. Hidaka M (1985) Biol. Bull. 168:350
139. Mariscal RN (1974) in: Muscatine L, Lenhoff HM (eds) Coelenterate biology: reviews and new perspectives. Academic Press, London, New York, pp 129–178
140. Blanquet RS (1977) in: Bernheimer AW (ed) Perspectives in toxinology. Wiley, New York p 150
141. Mariscal RN (1984) in: Bereiter-Hahn J, Matoltsy AG, Richards KS (eds) Biology of the Integument. Springer-Verlag, Berlin Heidelberg, pp 57–68 (Invertebrates, vol 1)
142. Chornesky EA (1983) Biol. Bull. 165:569
143. Francis L (1973) Biol. Bull. 144:73
144. Francis L (1976) Biol. Bull. 150:361
145. Sebens KP (1984) Biol. Bull. 166:457
146. Dyrynda PEJ (1986) Phil. Trans. R. Soc. Lond. B 313:27
147. Stebbing ARD (1973) in: Larwood GP (ed) Living and fossil bryozoa. Academic Press London, pp. 173–183
148. Goodbody I (1962) Biol. Bull. 122:40
149. Davis AR, unpublished data
150. Bak RPM, Sybesma J, Van Duyl FC (1981) Mar. Ecol. Prog. Ser. 6:43
151. Stocker LJ, Berquist PR (1986) J. Exp. Mar. Biol. Ecol. 102:1
152. Davis AR (1988) Mar. Ecol. Prog. Ser., 51:107
153. Buss LW (1979) in: Larwood G. Rosen BR (eds) Biology and systematics of colonial organisms. Academic Press, London, New York, pp 459–497
154. Jackson JBC (1979) In: Larwood G. Rosen BR (eds) Biology and systematics of colonial organisms. Academic Press, London, New York, pp 499–555
155. Tursch B, Braekman JC, Daloze D, Kaisin M (1978) in: Scheuer PJ (ed) Marine natural products. Academic Press, New York, p 247 (vol 2)
156. Weinheimer AJ, Metzner EK, Mole ML Jr (1973) Tetrahedron 29:3135
157. Ciereszko LS, Mizelle JW, Schmidt RW (1973) in: Worthen LR (ed) Food-Drugs from the sea proceedings, 1972. Marine Technology Society Publications, Washington DC, pp 177–180
158. Sullivan B, Faulkner DJ, Webb L (1983) Science 221:1175
159. Schmitz FJ, Lorance ED (1971) J. Org. Chem. 36:719
160. Gucinski H, Baier RE, Meyer AE, Fornalik MS, King RW (1984) in: Investigation of the Initial Events in Biofouling. Prepared for ONR Oceanic Biology Program, ONR Contract No. N00014-81-C-0671
161. White D (1988) in: Thompson MF, Sarojini R, Nagabhushanam R (eds) Marine biodeterioration. Oxford and IBH Publishing Co., New Delhi
162. Costerton JW, Geesey GG, Cheng KJ (1978) Sci. Amer. 238:86
163. Sutherland IW (1980) in: Berkeley RCW, Lynch JM, Melling J, Rutter PR, Vincent B (eds) Microbial Adhesion to Surfaces. Ellis Horwood Ltd, Chichester
164. Sutherland IW (1983) CRC Crit. Rev. Microbiol. 10:173
165. Fletcher M, Marshall KC (1988) in: Thompson MFD, Sarojini R, Nagabhushanam R (eds) Marine biodeterioration. Oxford and IBH Publishing Co., New Delhi
166. Mitchell R, Maki JS (1988) in: Thompson MF, Sarojini R, Nagabhushanam R (eds) Marine biodeterioration. Oxford and IBH Publishing Co., New Delhi, pp 489–497
167. Muller WA (1973) in: Tokidka T, Nishimura S (eds) Proceedings of the 2nd International Symposium on Cnidaria. Publ. Seto Mar. Biol. Lab. pp 195–208
168. Cameron RA, Hinegardner RT (1974) Biol. Bull. 146:335
169. Neumann R (1979) Mar. Ecol. Prog. Ser. 1:21
170. Mihm JW, Banta WC, Loeb GI (1981) J. Exp. Mar. Biol. Ecol. 54:167
171. Brancato MS, Woollacott RM (1982) Mar. Biol. 71:51

172. Kirchman D, Graham S, Reisch D, Mitchell R (1982) J. Exp. Mar. Biol. Ecol. 56:153
173. Maki JS, Rittschof D, Costlow JD, Mitchell R (1988) Mar. Biol. 97:199
174. Hudon C, Bourget E, Legendre P (1983) Can. J. Fish. Aquat. Sci. 40:1186
175. Maki JS, Mitchell R (1985) Bull. Mar. Sci. 37:675
176. Meadows PS, Williams GB (1963) Nature 198:610
177. Crisp DJ (1984) in: Costlow JD, Tipper RC (eds) Marine biodeterioration: an interdisciplinary study. Naval Institute Press, Annapolis, Maryland, pp 103–126
178. Marshall KC, Stout R, Mitchell R (1971) J. Gen. Microbiol. 68:337
179. DiSalvo LH (1973) Limnol. Oceanogr. 18:165
180. Corpe WA (1970) in: Acker RF, Brown BF, DePalma JR, Iverson WP (eds) Proceedings of the 3rd International congress on marine corrosion and fouling, Evanston, IL Northwestern Univ. Press, pp. 598–609
181. Lynch JM, Poole NJ (1979) Microbial ecology: a conceptual approach. Blackwell, London
182. Bak RPM, Borsboom JLA (1984) Oecologia 63:194
183. Palumbi SR (1985) Am. Nat. 126:267
184. Littler MM, Littler DS, Taylor PR (1987) J. Exp. Mar. Biol. Ecol. 105:107
185. Todd CD, Turner SJ (1988) J. Exp. Mar. Biol. Ecol. 115:113
186. Harvell DC (1986) Am. Nat. 128:810
187. Vance RR (1983) Ecology 60:537
188. Nelson WG (1979) Estuaries 2:213
189. Al-Ogily SM, Knight-Jones EW (1977) Nature 265:728
190. Ryland JS, Stebbing ARD (1971) in: Crisp DJ (ed) Proceedings of the 4th European Marine Biology Symposium, Cambridge, Cambridge University Press, pp 105–126
191. Stebbing ARD (1971) J. Mar. Biol. Ass. UK 51:283
192. Feeny PP (1976) Rec. Adv. Phytochem. 10:1
193. Steinberg PD (1984) Science 223:405
194. Paul VJ, Fenical W (1987) in: Scheuer PJ (ed) Bioorganic Marine Chemistry, Vol 1, Springer-Verlag Berlin Heidelberg New York London Paris Tokyo, pp 1–29
195. Paul VJ, Van Alstyne KL (1988) Coral Reefs 6:263
196. Rinehart Jr KL, Kobayashi J, Harbour GC, Gilmore J, Mascal M, Holt TG, Shield LS, Lafargue F (1987) J. Am. Chem. Soc. 109:3378
197. Kinzer KF, Cardellina JH (1987) Tetrahedron Lett. 28:925
198. Munroe MHG, Luibrand RT, Blunt JW (1987) in: Scheuer PJ (ed) Bioorganic Marine Chemistry, Springer-Verlag, Berlin Heidelberg New York, pp 93–176
199. Morse DE, Hooker N, Duncan H, Jensen L (1979) Science 204:407
200. Morse ANC, Morse DE (1984) J. Exp. Mar. Biol. Ecol. 75:191
201. Morse DE (1985) Bull. Mar. Sci. 37:697
202. Paul VJ, Van Alstyne KL (1987) The Western Society of Naturalists, 68th Annual meeting, abstract only
203. Van Alstyne KL (1988) Ecology 69:655
204. Littler MM, Taylor PR, Littler DS (1986) Coral Reefs 5:63
205. Gould SJ (1988) Nat. Hist. 97:12

Feeding Attractants and Stimulants for Marine Gastropods

Kanzo Sakata[1]

Contents

Abstract

Feeding is essential for organisms, and many biological studies have revealed that chemoreception and chemotaxis are involved in feeding behavior. This paper reviews recent studies of gastropod chemoreception in feeding with particular emphasis on the bioorganic chemical bases involved.

1 Introduction

Many biological studies have revealed that gastropods, in common with many other animals, can detect chemical changes in the environment. These changes can be associated with metamorphosis [1, 2], settling [3, 4] and particular behavior such as feeding [5–8], homing [5, 7], mating [5, 7] and predator escape [5, 7] etc. In gastropods chemical reception is considered to be the principal method for distant detection of food [5]. As feeding is essential for all living organisms, selection and ingestion of appropriate foods can be expected to have an adaptive or evolutionary basis. The following list of chemical substances concerned with chemoreception in various kinds of feeding behavior is based on definitions by Mackie [9] and Lindstedt [10]. In all cases, the animal responds to stimulus involving a chemical gradient.

1. Arrestant...........orienting toward the apparent source, possibly over long distances.

1 Research Laboratory of Marine Biological Science, Faculty of Agriculture, Shizuoka University, Mochimune, Shizuoka 421-01, Japan.

Bioorganic Marine Chemistry, Volume 3

2. Attractant ....locomotion ceases close to the apparent source.
3. Repellent...........orientation away from the apparent source.
4. Incitant ....evocation of the initiation of feeding (tasting).
5. Suppressant.......inhibition of feeding initiation.
6. Stimulant ....promotion of ingestion and continuation of feeding.
7. Deterrent...........prevention of continued feeding or accelerated termination of feeding.

Recognition of each active substances is actually quite difficult. Feeding arrestants can be involved in the response to attractants in some cases. Feeding incitants, suppressants and stimulants are also difficult to distinguish unless appropriate bioassay procedures are used.

Recent neurophysiological studies have revealed that the chemoreceptors for the chemical stimuli are located in the cephalic tentacles, rhinophores, oral veil and osphradium [7, 11, 12]. Few chemical studies, however, have been related to these neurophysiological observations.

Invertebrates belonging to the phylum Mollusca are represented by about 110,000 species, which places them second only to the total insect species. Their distribution is very wide involving sea, river, lake, and land habitats. Eightyfour percent of Mollusca are gastropods, to which economically important shellfish like the abalone, *Haliotis* spp., and the turban shell, *Turbo* spp., belong. Fortyeight percent of the gastropods are found in the sea, 13% in freshwater, and 39% on land. Gastropods, both carnivores and herbivores, generally search for prey in contrast to bivalves which are plankton feeders. The class Gastropoda is composed of three subclasses, Prosobranchia, Opisthobranchia, and Pulmonata, as shown in Table 1. Many marine gastropods like the abalone, *Haliotis* spp., and the cone snails, *Conus* spp., are included in Prosobranchia. Opisthobranchia have very poorly developed shells and continuously expose their colorful soft bodies. Special

Table 1. Classification of mollusca

Mollusca	
Polyplacophora	
Aplacophora	
Monoplacophora	
Gastropoda	
Prosobranchia	cone snails, *Conus* spp., abalones, *Haliotis* spp., turban shells, *Turbo* spp., etc. having well-developed shells, gills forwards against hearts and opercula
Opisthobranchia	sea hares, *Aplysia* spp., etc. having poorly developed shells and gills backward against hearts
Pulmonata	*Acathina* spp., *Biomphalaria* spp., etc. having lungs instead of gills and no opercula
Scaphopoda	
Bivalvia	
Cephalopoda	

attention has been paid to one of the orders, Nudibranchia; its chemical ecology has quite recently been reviewed by Karuso [13] emphasizing their chemical defense against predators. Pulmonata have lungs instead of gills as an adaptation to land habitats, where many of them are considered to be pests for crops and gardens.

Chemical studies on the feeding behavior of gastropods may be expected to contribute to advances in the culture of commercially valuable gastropods like abalones and to the ability to control snail pests.

2 Marine Gastropods

2.1 Carnivores

Observations of gastropod behaviors in nature suggest that chemoreception is the most important means of food detection among a wide variety of carrion-eating species [5]. The mud snail *Nassarius* sp. is reported to respond to substances diffusing from dead fish and crustaceans by repeatedly extending its proboscis. Carr found that lactic acid and glycine, which were the main constituents of the active fraction obtains from a shrimp extract, stimulated the response at 5×10^{-1} and 10^{-3} M, respectively [6]. The proboscis response seems to involve the combined or possible synergistic effects of a number of compounds because the response to crude extracts is greater than the response to any single compound. The same phenomenon has been observed in feeding attractants for fish [14–16] and crustaceans [17].

Nassarius spp. are reported to respond to the body fluid of the oyster and blood serum of human beings. Glycoprotein and albumin have been isolated and identified as the active constituents of these [6].

Most of the 400 species of cone shells, *Conus* spp., are found in tropical waters. They feed exclusively on live fish or gastropods. Their feeding behavior is of interest [6]. For example, *Conus magus*, as other fish-hunting snails, lures its prey with a brightly colored proboscis writhing above the sand, while the rest of the body lies buried. When the fish tries to bite the worm-like proboscis, it is stung in the mouth with the radular tooth and paralytic peptides called conotoxins [18] are injected. This feeding behavior was shown to be very sensitive to the presence of live prey as the response occurs within a few seconds to live fish, but not to dead fish [7]. The basis for this chemical stimulus has not yet been elucidated.

Oyster drills, *Urosalpinx cinerea*, have the well-documented ability to locate intact prey such as oysters and barnacles from a distance. This response increases when young oysters are involved. Oxaloacetic acid, which is considered to be one of the final products in shell formation, is one attractant in this response [5–7]. The actual active principle, however, still remains unknown. Recently Rittschof et al. [19] isolated attractants for newly hatched snails from seawater in which the barnacle *Balanus eburneus* has been kept. These attractants were relatively large, at least partially proteinaceous molecules (>1000 but $<100{,}000$ Da), but their structures have not been elucidated. Complex chemotaxis and chemoreception are involved in this prey-predator behavior and Rittschof et al. are examining such multifunctional communication systems in the marine environment [20].

The gastropod *Thias* collected in the field exhibits a strong dietary preference for the blue mussel, *Mytilus edulis*; this did not change during 4–9 weeks, when the food supply was changed to *M. californica*, a very closely related species to *M. edulis*. Specific chemical food cues seem to be very important for such dietary preferences of newly born *Achatina* sp. could be modified by experience with different foods.

Davis and coworkers [22] studied sensory reception in the mollusc *Pleurobrancheae californica*, which feed on the sea anemone *Corynactis* sp. [22]. The whole animal consistently responded to 10 of 18 amino acids tested, including alanine[1], glycine, etc. When the rhinophore nerve was surgically removed, strong centripetal responses were observed for these two amino acids, but not to glutamic acid which is an attractant for most other aquatic animals such as fish [14–16]. Individual amino acid was generally less effective than food stimuli (*Corynactis* homogenate) in eliciting feeding behavior of the gastropod, thus suggesting synergistic effects for these amino acids as in the case of fish [14–16] and crustaceans [17].

Although the importance of chemoreception in searching for food by predators has been demonstrated in many cases, only few chemical studies have been conducted to determine what kind of compounds are involved. Perhaps each animal has developed an ability to detect different chemical signals that are specifically related to the main food material available throughout their long evolutionary development. This chemical-ecological approach to studies of prey-predator interaction has been followed for abalone [33–35].

2.2 Herbivores

Among gastropods, the abalone, *Haliotis* spp., provide a popular seafood for gourmets. Abalones generally require several years to reach market-size and, partly for this reason, excessive harvesting has led to population depletion. Recently, artificial spawning, fertilization and hatching techniques for abalone [4, 23] have become well established and more than ten million young abalone or seedlings are now produced every year at local Fish Farming Centers all over Japan [24, 25]. These seedlings have been released into the sea to increase local production and the beneficial effects of this restocking are eventually being felt.

In culturing juvenile abalone, the main natural foods are attached diatoms at an early stage, followed by the green alga *Ulva pertusa* and the brown alga *Eisenia bicylis*. It generally takes more than a year and half to obtain juvenile abalone with a shell length of about 20–25 mm which is recognized as the minimum size for release into the sea. This minimum size is important as the survival rate of the liberated abalone depends largely on their size at the time of release [24]. The smaller they are, the more they are eaten by predators. Thus, recently developed artificial diets for juvenile abalone were used to produce larger "seedlings" as well as

---

[1] Unless otherwise stated, amino acids are L-forms

shorten the culture period. The artificial diets include the addition of feeding stimulants as well as nutrients, where the current practice is to add powdered seaweed as a feeding stimulant. Few chemical studies, however, have been made of the feeding stimulants that form part of the feeding behavior of such important gastropods as the abalone species of *Haliotis*.

In the feeding behavior of the abalone *Haliotis discus*, the importance of olfaction was suggested. Kikuchi et al. [26] observed that the abalone *H. discus hannai* fed on most algae, but showed feeding preference for some kinds of brown and red algae. Uki et al. [27, 28] demonstrated that brown algae including *Undaria pinnatifida* and *Eisenia bicyclis* are the most suitable food for abalone growth. In contrast, Australian abalone were reported to feed preferentially on red algae and a particular species, *H. laevigata*, on sea grass [29].

Harada et al. [30] studied the attractive effect of seaweeds using a test aquarium similar to that used for the feeding attractant study of eel by Konosu et al. [15]. Fifty or one hundred individuals of the young abalones, *H. discus*, were introduced into a test aquarium and their behavior was observed for 20 min. Some of them enter the compartment where a test sample was placed, but all of them never remain there. Some of them migrate out of the compartment. The number of the test animals in each compartment were counted every two minutes to make a time-course of their response. By application of logistic curve analysis to the results, they found that some species of green and red algae were effective in attracting the abalone, but brown algae generally tended to attract them more [30]. The behavioral responses of the young abalone to some of the algal constituents, protein and lipid fractions were revealed to be involved in this attraction [31]. Nucleic acid related compounds such as adenine, adenosine-5'-monophosphate (AMP), etc. [32] were found to be attractants for the young abalone, too. A synergistic effect was also observed.

The expected shortage of fishery resources will soon emphasize the need to establish culturing methods for commercially useful gastropods including abalone. For this purpose, the use of phagostimulants or feeding stimulants is likely to be more important than feeding attractants for seedling production and culture. These considerations prompted a chemical study on the feeding behavior of the abalone *H. discus* [33–35, 55] and some of the new interesting results are related below.

We kept young abalone, *H. discus*, with shell length about 1.5–2.5 cm in an aquarium and observed their feeding behaviors. They are nocturnal and hide beneath the shelter in the test aquarium during daytime. Their food searching begins after sunset. Careful observation of their feeding behavior led to establishment of a simple and reliable bioassay procedure for chemical study of their feeding preferences [33]. A glass plate (5 × 20 or 10 × 20 cm coated with crystalline cellulose, Avicel SF and dried) with several sample zones with a diameter of 2.5 cm drawn with a pair of compasses was used as a test plate. Each sample solution was absorbed on a sample zone. The test animals were starved at least a day before the assay. The Avicel plate ready for the assay was placed at the bottom of the test aquarium after sunset. Figure 1 shows an overview of an experimental aquarium during the bioassay. The test plate was taken out the next morning. Figure 2 shows

Fig. 1. An overview of the experimental aquarium of *H. discus* during the bioassay. During the daytime abalones were hiding under the shelter. This photograph was taken at 9.30 p.m. The abalones were moving around searching for food. A sample zone has just been bitten off by an abalone

typical assay results. Assessment of the feeding activity was easily achieved. The results of this bioassay method indicated an interesting feeding behavior of *Haliotis* sp. They seem to creep on the test plates to search for foods leaving a pair of their characteristic feeding traces here and there and bite off the Avicel when they come across a sample zone where their feeding stimulants are absorbed. This assay method is applicable to identification of feeding stimulants rather than attractants.

Among the methanol extracts of several seaweeds subjected to this bioassay, the brown algae, *Undaria pinnatifida* and *Padina arborescens*, were judged to be the most active [33]. Application of this assay procedure to the extracts of *U. pinnatifida* allowed us to isolate two groups of active principles [35]. Spectroscopic and chemical studies revealed that they are digalactosyldiacylglycerol (DGDG) and phosphatidylcholine (PC) having highly unsaturated fatty acid moieties composed of C_{14}–C_{22} chains [34, 35]. This was the first report to have clarified the feeding stimulant activity of such glycerolipids as DGDG and PC, which are rather lipophilic compounds. All the feeding attractants and stimulants of marine organisms elucidated so far are all water-soluble substances.

Sea hares belonging to Opisthobranchia have very poorly developed shells and expose their body all the time. Frings and Frings [36] reported that the Hawaiian sea hare, *Aplysia juliana*, is almost monophagous and feed only on the green alga *Ulva lactuca* and they may nibble on other algae, but never eat enough to grow or survive for long. They also observed that the water extracts of the fresh *Ulva* sp. elicited characteristic phagostimulatory responses in the sea hare. Based on the typical opening mouth response and electro-physiological experiments, amino acid constituents of natural foods were suggested to be food attractants for this species [37]. Recently Carefoot found that addition of water extracts of *Ulva* sp. to his artificial diet for *Aplysia* sp. was necessary [38] and other constituents such as

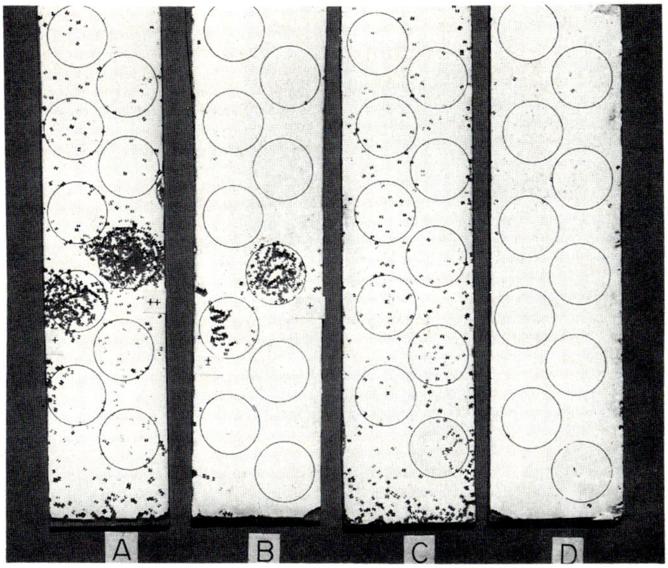

Fig. 2. Typical assay results obtained by using young abalones *H. discus*. Aliquot of each fraction from the Sephadex LH-20 column chromatography of the MeOH extracts of *E. bicyclis* were absorbed in each sample zone (2.5 cm in diameter) on the Avicel plate (0.25 mm thick, 5 × 10 cm). After evaporation of the solvent each test-plate was placed, after sunset, at the bottom of each test aquarium which contains 10–20 young abalones (2–2.5 cm in shell length). Plates A and C were placed together in a test aquarium and B and D in another. Next morning, the plates were taken out of the aquaria and the activity was judged as + +, +, and −, according to the ratio of the feeding traces (a pair of characteristic feeding traces) inside the sample zone to those outside. The abalones in the test aquarium for plates A and C seem to be more starved than those in the other. Characteristic feeding traces were left all over the test plates, which means that they search for foods by biting off the Avicel here and there and keep biting when they reach the sample zone in which their feeding stimulants were absorbed. This assay method is clearly suitable for testing feeding stimulants

starch, maltose, etc. elicited phagostimulatory responses [39, 40]. None of the previous studies had identified the chemical basis for the real phagostimulants in the algae, which was indicated by the bioassay results of algal extracts.

When checking the feeding behavior of several young sea hares *A. juliana*, we observed feeding preference for filter paper on which an ether extract of *Ulva* sp. had been absorbed. Several trials have allowed us to establish a simple and reliable bioassay procedure for feeding stimulants for the young sea hare *A. juliana* [41] based on this observation.

Young sea hares, *A. juliana*, were kept in a test aquarium (32 × 61 × 18 cm) with a drain to maintain a water depth of 5 cm. Filter papers (5.5 cm in diameter) treated with sample solution and provided with sinkers were placed at the bottom of the test aquarium and the assay was started. The sea hares moved around; some

Fig. 3. An overview of a test aquarium for *A. juliana* during the bioassay. Ten test animals were shifted along the corner of the bottom. Filter papers (5.5 cm in diameter) absorbed aliquot of fractions from silica gel column chromatography of the extracts of a green alga *U. pertusa* were placed with pebbles or sinkers on the bottom of the aquarium. This photograph was taken 20 minutes after the beginning of the assay

of them showed their typical feeding stimulatory responses [37]; and some of them reached the filter paper and started eating it. A fraction is obviously very active (Fig. 3). The numbers of individuals at each filter paper were counted after 10, 20 and 30 min. Judgment of activity was based on the total number of sea hares aggregated on each filter paper.

Application of this assay procedure to one of the most active algal extracts, *Ulva pertusa*, allowed us to isolate active principles, which were identified as DGDG and 1,2-diacylglyceryl-4'-*O*-(*N*,*N*,*N*-trimethyl)-homoserine (DGTH)[1] [42]. It is rather interesting to note that DGDG which was isolated as a feeding stimulant for young abalone [34, 35] is also active in the sea hare assay. DGTH is an interesting glycerolipid recently isolated from a limited number of plant species such as pteridophytes and green algae [43–51] and from a fungus [52]. Methyl esters liberated from DGTH by methanolysis with methanolic KOH were identified as a mixture of fatty acid methyl esters (C_{14}–C_{22}) [42].

The turban shell ("sazae" in Japanese), *Turbo cornutus*, and the top shell ("bateria" in Japanese) *Omphalius pfeifferi*, are both marine herbivorous gastropods of the subclass Prosobranchia whose delicate taste is highly esteemed by the Japanses. Annual yields of the turban shell harvest have been decreasing by overcapturing and some trials for its seedling production have recently been started in Japan.

Some biological studies on the turban shell clarified their feeding behavior: i) they are nocturnal and begin food-searching after sunset; ii) they feed on brown

[1] Two abbreviations, DGTH and DGTS, have been used for this compound. We prefer DGTH, because DGTS reminds us of a serine derivative rather than a homoserine one.

algae such as *Eisenia bicyclis* and *Undaria pinnatifida* and red algae, *Gelidium* spp., and grow well on the brown algae *E. bicyclis* [53].

No report has been found on the feeding behavior of the top shell, *O. pfeifferi*. There are no reports of chemical studies on the feeding behavior of these herbivorous gastropods.

Recently we found that the "Avicel plate method" [34] was applicable to chemical studies on feeding stimulants for these gastropods [54]. Assay results were more reproducible in the case of *Omphalius* than *Turbo*. Typical feeding traces were found on the plate and judgment of the activity of each sample was performed as in the case of the abalone [34]. Interestingly the feeding traces left by *Turbo* have the same shape as those made by *Haliotis*, but those by *Omphalius* are clearly different [54].

Finally we have established bioassay procedures for phagostimulants for 4 different species of marine herbivorous gastropods. Several methanolic algal extracts were subjected to the assay. The extract of the green alga *Ulva pertusa* was the most active for all the test animals. Accordingly, *U. pertusa* was selected for the isolation of the phagostimulants contained in the alga. Chromatographic purification yielded several glycerolipids, DGDG, DGTH, 1-monoacylglyceryl-4'-*O*-(*N*,*N*,*N*-trimethyl)-homoserine (MGTH), 6-sulfoquinovosyldiacylglycerol (SQDG) [55] as phagostimulants (Fig. 4). Although SQDG is present in brown algae, *E. bicyclis* and *U. pinnatifida*, it was not isolated as a feeding stimulant for the abalone, *H. discus*, in our previous study [35], but for the turban shell, *T. cornutus*, and the top shell, *O. pfeifferi*. This means that food preference of these herbivorous gastropods seems to be similar, but slightly different from that of *Haliotis*.

These results thus confirm the activity of DGDG as a feeding stimulant together with PC for the young abalone *H. discus* and for the sea hare *A. juliana*. The phagostimulatory activity of the organic extracts of this green alga can, therefore, be attributed to the glycerolipids DGDG, DGTH, MGTH, and SQDG.

One very interesting point is that while DGDG, DGTH, MGTH, and PC were found to be effective for all the gastropods being tested, SQDG was active only for *T. cornutus* and *O. pfeifferi*. The abalone *H. discus* did not response to SQDG even

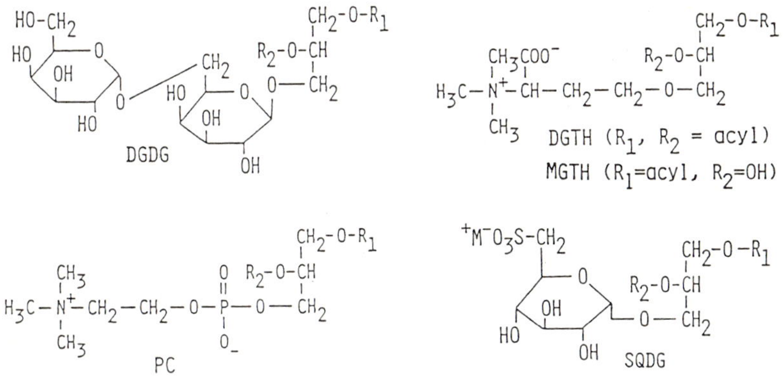

Fig. 4. Glycerolipids as phagostimulants

Table 2. Feeding-stimulation activity of the glycerolipids isolated from the alga *Ulva pertusa* [55]

	Haliotis discus	*Turbo cornutus*	*Omphalius pfeifferi*	*Aplysia juliana*
DGDG	14–28 μg[1]	15–25	18–23	800[2]
DGTH	10–20	24	<10	100
MGTH	23–50	25–50	<23	
SQDG	>300	20–40	<20	>1000
PC[3]	<10	15	<15	<50

[1] MED (minimum effective dose) value expresses minimum sample amount/sample zone necessary to show an activity (+). These values are preliminary.
[2] In the case of *Aplysia* minimum sample amount/filter paper (Toyo filter paper No. 2, 5.5 cm in diameter).
[3] Isolated from the brown alga *Undaria pinnatifida*.

at doses as high as 300 μg (Table 2). The young abalone was, however, very responsive to amounts of DGDG and DGTH as low as 20–30 μg.

Catalytically hydrogenated PC showed the same activity as PC for the young abalone [35], thereby suggesting that unsaturation in the acyl groups is not related to activity. Further structure-activity relationship is now under investigation.

The wide occurrence of DGDG and PC in the plant kingdom strongly suggests that marine herbivorous gastropods could feed not only on any algae but also on terrestrial plants. The abalone *H. discus hannai* is, however, reported to show quite different feeding preferences. Some algal species are not eaten by them, even though they are not so tough [28]. Some suppressant (feeding deterrent) may be present in these particular species of algae. Our new bioassay method "Avicel plate method" [33] is now being applied to study this interesting phenomenon.

3 Concluding Remarks

As mentioned in the introduction (Sect. 1) to this chapter, some species of gastropods live in freshwater and on land. This review is concerned with chemoreception in marine gastropods, but I would like to describe very briefly the chemotaxis and chemoreception in freshwater gastropods, because there seems to be no difference between chemoreception in the sea and in freshwater.

Much evidence has been accumulated to show that pulmonate snails, like other gastropod molluscs, use their chemical sense as the principal modality for the detection of various targets including food [7, 56]. *Biomphalaria glabrata* is a freshwater pulmonate snail and one of the snail hosts of *Schistosoma* spp. which cause serious schistosomiasis in Africa. Uhazy et al. established a very simple bioassay method applicable to chemical studies on feeding attractants for the juvenile snail and isolated glutamic acid (Glu) and proline (Pro) as the main feeding attractants from the homogenates of lettuce to which the animals were known to be attracted[57]. Interestingly, these workers observed that other amino acids such as glycine, threonine, and valine clearly showed repellent activity against the test animals. Those chemicals may be related to prey-predator interaction.

Thomas et al. established a simple bioassay method [58] which can tell attractant, arrestant, or repellent activity for 9 species of pulmonate snails, *B. glabrata*, *Buliness globosus*, *Lymnasa natalensis*, *Physella acuta*, etc., using a so-called diffusion olfactometer [59]. Very extensive screening [58–64] clarified feeding stimulants as well as attractants by using pure chemicals which are considered to be constituents of their natural food material together with their analogs. They found 10 amino acids like tryptophan, serine, asparagine (Asn), 2,4-diaminobutyric acid, ornithine, etc. out of 30 amino acids showed attractive activity for the young animal [58]. Interestingly, L- and D-Asn showed no difference in activity at 4.5×10^{-4} M. Adult snails were attracted by aspartic acid, Glu, Pro, hydroxy-Pro, and propionic acid. Propionic acid, which is known to be excreted from a gland present in the foot, was more active than any amino acid [60]; it may be a kind of pheromone or kairomone rather than a feeding attractant. Adult snails were also attracted by both L- and D-Asp, which means their receptor cannot distinguish absolute stereostructures of amino acids. This is in sharp contrast to fish and crustaceans, which exclusively detect only L-amino acids as their feeding attractants [14–17]. Among 24 sugars and their derivatives tested, maltose and maltotriose were active [63]. The latter was active even at such low concentrations as 5×10^{-6}–10^{-7} M.

Recently Thomas et al. [64] screened feeding stimulants for adult snails using an apparatus which allows to observe the action of the muscle in their mouth. Only five kinds of carbohydrates (xylose, glucose, mannose, maltose, and maltotriose) were active at 10^{-2} M among 41 kinds of sugars. Juvenile snails were found to be stimulated by more kinds of sugars than adult ones.

Chemical studies on feeding attractants for the land snail *Acathina fulica* [65, 66] and the field slug *Deroceras reticulatum* [67] have recently been conducted.

As seen in the case of fish [14–16] and crustaceans [17], amino acids [5, 6, 7, 22], proteinaceous compounds [6, 19, 31], and nucleic acid-related compounds [32] have been shown to be involved in chemotaxis in gastropods, also. They are all water-soluble compounds. The feeding stimulants for herbivorous gastropods which we studied were glycerolipids like DGDG, PC, DGTH, SQDG, etc. [34, 35, 42, 55, 68]. These compouds are the first examples of lipophilic feeding stimulants for marine organisms. All feeding attractants and stimulants for fish so far elucidated are water-soluble compounds, such as amino acids, nucleosides, etc. [14–17].

The sea urchin is another marine herbivore that is an important seafood in Japan. Behavioral studies have revealed their feeding preferences for some particular species of algae [69, 70]. Quite recently, Machiguchi and we found that the "Avicel plate method" is well applicable to feeding stimulant studies of the sea urchin, *Strongylocentrotus intermedius* [71]. It was revealed that glycerolipids, as e.g. DGDG, PC, SQDG, were also their strong feeding stimulants [72]. The sea urchin left their typical feeding traces like star-marks in the sample zone where feeding stimulants had been absorbed. Judgment of the activity can be made in exactly the same manner as in the cases of abalone. It is quite interesting that those glycerolipids also stimulate the feeding of sea urchins.

It is worthwhile to note that the sea urchin *S. intermedius* is now easily cultured by just feeding it the terrestrial plant, *Reynoutria sachalinensis* ("ooitadori" in Japanese). This technique has been extended to seedling production of the sea urchin in the Hokkaido district of Japan [73].

Chemoreception and chemotaxis involved in feeding behavior of marine gastropods have been reviewed. Here the author would like to point out recent biochemical studies which have revealed the involvement of chemicals in spawning, metamorphosis and settlement in gastropods. Raising water temperature and UV-irradiation were well-known exogenous factors to induce maturation and spawning in *Haliotis* spp. [74, 75] and are actually applied for their year-round seedling production [75]. Morse et al. [4, 76] reported that addition of hydrogen peroxide ($\sim$ 5 mM) to seawater caused synchronous spawning in the Californian abalone, *H. rufescens*, and a kind of blue mussel, *Mytilus californianus*, as well. They suggested that the H_2O_2 activation of prostaglandin endoperoxide synthesis might account for the H_2O_2-dependent induction of spawning. Matsutani et al. [77, 78] recently found that one of neurotransmitters, serotonin (5-hydroxytryptamine), also induced spawning in the scallop, *Patinopecten yessoensis*, and the oyster, *Crassostrea gigas*.

Planktonic larvae of the abalone, *H. rufescens*, were known to exhibit substrate-specific settlement in response to the crustose coralline red algae *Lithothamnium* spp. and *Lithophyllum* spp. Morse et al. [79] found the γ-aminobutyric acid (GABA) was active in inducing behavioral metamorphosis at concentrations as low as $\sim 10^{-7}$ M, with half-maximal effectiveness at $< 10^{-6}$ M. GABA was actually found in high concentration in the red algae extracts. They found that the algal pigments, phycoerythrobilin and phycoerythrin, [79] and the red algal phycolipoproteins [80] were also active.

Recently, Nomura and coworkers [81] examined effectiveness of several neurotransmitters such as GABA, serotonin, glutamic acid, etc. in induction of larval settlement and metamorphosis of the abalone, *H. discus hannai*, and found that these chemicals depressed ciliary movement of the larval velum to make the larvae fall down to the bottom, crawl shortly and die. None of them induced larval metamorphosis in contrast to the case of H. *rufescence* mentioned above. Seki et al. [82] revealed the real inducers for the settlement and metamorphosis of the abalone as the mucus left behind by the abalones. Interestingly, they observed that the mucus secreted by artificially stimulating the abalone is not so active as the one left by them while they were crawling on the plate by themselves. As mentioned above, only a part of the larval settlement and metamorphosis have been chemically clarified yet.

For a long time many biological studies suggested that marine organisms are controlled by chemical signals. Only a limited number of chemical studies, however, have been conducted so far. Because of economic demands much work has been carried out on commercially valuable fish and crustaceans, but litte on other organisms. It will be necessary to extend out studies on chemoreception and chemotaxis in the marine ecosystem from a chemical and ecological point of view.

References

1. Burke RD (1983) Can. J. Zool. 61:1701
2. Yvin JC, Chevolot L, Chevolot-Magueur AM, Cochard JC (1985) J. Nat. Prod. 48:814
3. Nomura T, Matsutani T (1987) Bioactive substances in molluscs. In: Yasumoto T, Kamiya H (eds) Bioactive marine natural products, Vol. 65, Koseisha Koseikaku, Tokyo, pp 111-124 (Suisangaku Series vol 65)
4. Morse DE (1984) Aquaculture 39:263
5. Kohn AJ (1961) Am. Zoologist 1:291
6. Lenhoff HM, Lindstedt KJ (1974) Chemoreception in aquatic invertebrates with special emphasis on the feeding behavior of coelenterates. In: Grant PT, Mackie AM (eds) Chemoreception in Marine Organisms, Academic Press, London, p 147
7. Croll RP (1983) Biol. Rev. 58:293
8. Sakata K (1988) Feeding behavior stimulants of gastropods. In: Shibata S (ed.) Shinpen Seibutukassei Tennenn Busshitsu, Ishiyaku Shuppan, Tokyo, p 138
9. Mackie AM (1982) Identification of the gustatory feeding stimulants. In: Harada TJ (ed.) Chemoreception in fishes, Elsevier Sci. Pub, Amsterdam, p 275
10. Lindstedt KJ (1971) Comp. Biochem. Physiol. 39A:553
11. Chase R (1981) Comp. Biochem. Physiol. 70A:149
12. Horwitz IS, Senseman DM (1981) Chem. Senses 6:1
13. Karuso P (1988) Chemical ecology of the nudibranchs. In: Scheuer PJ (ed.) Bioorganic Marine Chemistry, Vol. 1, Springer-Verlag, Berlin, p 31
14. Hara TJ (1982) (ed.) Chemoreception in Fishes. Elsevier Sci. Pub. Co, Amsterdam
15. The Japanese Society of Scientific Fisheries (1981) (ed) Chemical Sense of fish and feeding stimulants. Koseisha Koseikaku, Tokyo
16. Ina K (1988) Feeding attractants and stimulants of fish. In: Shibata S (ed) Shinpen Seibutukassei Tennenn Busshitsu. Ishiyaku Shuppan, Tokyo, p 131
17. Carr WES, Derby CD (1986) J. Chem. Ecol. 12:989
18. Cruz LJ, Gray WR, Yoshikami D, Olivera BM (1985) J. Toxicol.-Toxin Rev 4:107
19. Rittschof D, Shepherd R, Williams LG (1984) J. Chem. Ecol. 10:63
20. Rittschof D (1985) Amer. Malacolog. Bull. Special Ed. No. 1:111
21. Croll RP, Chase R (1980) J. Comp. Phys. 136:267
22. Bicker G, Davis WJ, Matera EM, Kovac MP, StormGipson DJ (1982) J. Comp. Physiol. 149A:221
23. Uki N, Kikuchi S (1984) Aquaculture 39:247
24. Uki N (1979) Abalone Culture in Japan. In: Sindermann CJ (ed) Proc Ninth and Tenth US-Japan Meetings on Aquaculture, p 83
25. Uki N (1986) Kagaku to Seibutsu 24:495
26. Kikuchi S, Sakurai Y, Sasaki M, Itoh T (1967) Bull. Tohoku Fish. Res. Lab. 27:93
27. Uki N (1981) Bull. Tohoku Fish. Res. Lab. 42:19
28. Uki N, Sugiura M, Watanabe T (1986) Bull. Japan. Soc. Sci. Fish 52:257
29. Shepherd SA (1975) Australian Fisheries No.1:12
30. Harada K, Kawasaki O (1982) Bull. Japan. Soc. Sci. Fish. 48:617
31. Harada K, Maruyama S, Nakano K (1984) Bull. Japan. Soc. Sci. Fish. 50:1541

32. Harada K (1986) Bull. Japan. Soc. Sci. Fish. 52:1961
33. Sakata K, Itoh T, Ina K (1984) Agric. Biol. Chem. 48:425
34. Sakata K, Ina K (1985) Agric. Biol. Chem. 47:2957
35. Sakata K, Ina K (1985) Bull. Japan. Soc. Sci. Fish. 51: 659
36. Frings H, Frings C (1965) Biol. Bull. Mar. Biol. Lab. (Woods Hole) 128:211
37. Jahan-Pawar B (1972) Am. Zoologist 12:525
38. Carefoot TH (1980) J. Exp. Mar. Biol. Ecol. 42:241
39. Carefoot TH (1981) Can. J. Zool. 59:445
40. Carefoot TH (1986) Mar. Biol. 68:207
41. Sakata K, Tsuge M, Ina K (1986) Marine. Biol. 91:509
42. Sakata K, Tsuge M, Kamiya Y, Ina K (1985) Agric. Biol. Chem. 49:1905
43. Brown AE, Elovson J (1974) Biochem 17:3476
44. Eichenberger W, Boschetti A (1978) FEBS Letters 88:201
45. Moseley KR, Thompson Jr. GA (1980) Plant Physiol. 65:260
46. Fried A, Tietz A, Ben-amotz A, Eichenberger W (1982) Biochim. Biophys. Acta 713:419
47. Janero DR, Barrnett R (1982) J. Lipid. Res. 23:307
48. Eichenberger W (1982) Plant Sci Lett 24:91
49. Evans RW, Kates M, Wood GW (1982) Chem. Phys. Lipids 31:331
50. Sato N, Furuya M (1983) Plant Cell Physiol. 24:1113
51. Sato N, Furuya M (1984) Phytochem. 23:1625
52. Yamada T, Nozawa Y (1972) Biochim. Biophys. Acta 574:433
53. Tamura T (1976) Senkai Zoshoku Gaku, Koseisha Koseikaku, Tokyo, p. 187
54. Sakata K, Sakura T, Kamiya Y, Ina K (1988) Nippon Suisan Gakkaishi, 54:1715
55. Sakata K, Sakura T, Ina K (1988) J. Chem. Ecol. 14:1405
56. Thomas JD (1982) Malacologia 22:81
57. Uhazy LS, Tanaka RD, MacInnis AJ (1978) Science 201:924
58. Thomas JD, Assefa B (1979) Comp. Biochem. Physiol. 63C:99
59. Thomas JD, Ofosu-Barko J, Patience RL (1983) Comp. Biochem. Physiol. 75C:57
60. Thomas JD, Assefa B, Cowley C, Ofosu-Barko J (1980) Comp. Biochem. Physiol. 66C:17
61. Sterry PR, Thomas JD, Patience RL (1983) Freshwater Biol. 13:465
62. Thomas JD, Ndifon GT, Ukoli FMA (1985) Comp. Biochem. Physiol. 82C:91
63. Thomas JD (1986) Comp. Biochem. Physiol. 83A:457
64. Thomas JD, Sterry PR, Jones H, Gubala M, Grealy BM (1986) Comp. Biochem. Physiol. 83A:461
65. Croll RP, Chase R (1980) J. Comp. Physiol. 136:267
66. Chase R (1982) J. Comp. Physiol. 148:225
67. Pickett JA, Stephenson JW (1980) J. Chem. Ecol. 6:435
68. Sakata K (1988) Feeding attractants and stimulants of gastropods. In: Shibata S (ed) Shinpen Seibutukassei Tennenn Busshitsu. Ishiyaku Shuppan, Tokyo, p 138
69. Machiguchi Y (1987) Bull. Hokkaido Reg. Fish. Res. Lab. No.51:33
70. Hay ME, Lee Jr. RR, Guieb RA (1987) J. Exp. Mar. Biol. Ecol. 96:147
71. Machiguchi Y, Sakata K (1987) Abstracts Autumn Meeting Jap. Soc. Sci. Fisheries, October 1987, Hakodate, p 167
72. Sakata K, Kato K, Ina K, Machiguchi Y (1989) Agric. Biol. Chem. 53:1457
73. Department of Mariculture, Hokkaido Cent. Fish, Exp. Sta. (1984) Month. Rep. of Hokkaido Fish. Exp. Sta. 41:270
74. Uki N, Kikuchi S (1982) Bull. Tohoku Reg. Fish. Res. Lab. 44:83
75. Uki N, Kikuchi S (1984) Aquaculture 39:247
76. Morse DE, Duncan H, Hooker N, Morse A (1976) Science 196:298
77. Matsutani T, Nomura T (1982) Mar. Biol. Lett. 3:353

78. Matsutani T, Nomura T (1986) Bull. Japan. Soc. Sci. Fish. 52:1589
79. Morse DE, Hooker N, Duncan H, Lloyd J (1979) Science 204:407
80. Morse ANC and Morse DE (1984) 75:191
81. Akashige S, Seki T, Kan-no H, Nomura T (1981) Bull. Tohoku Reg. Fish. Res. Lab. 43:37
82. Seki T, Kan-no H (1981) Bull. Tohoku Reg. Fish. Res. Lab. 43:29

Discovery and Development of Antineoplastic Agents from Natural Sources

Matthew Suffness, David J. Newman[a], and Kenneth Snader

Contents

Natural Products Branch, Developmental Therapeutics Program, Division of Cancer Treatment, National Cancer Institute, N.I.H. Bethesda, MD 20892, USA and [a]2026 14th Place, Vero Beach, FL 32960, USA

Bioorganic Marine Chemistry, Volume 3

Abstract

The main determinant of which compounds will be isolated from a natural source is the bioassay used to detect activity and to follow activity during fractionation. Suitable assays range from the very general cell cytotoxicity to highly specific interactions with receptors and enzymes; the choice of assay is based on the biological properties desired in the final compound, the hypothesis being tested, and the relationship to the more detailed secondary screens which will be subsequently used to select compounds for development. While highly specific mechanism-based or receptor-based assays miss compounds which exert their activity by other means, these assays have the major advantages of simplifying bioassay-directed isolation and simplifying identification of previously known compounds because of the smaller numbers of compounds they detect.

A key issue in development of active compounds is the ability to produce sufficient material for in vivo antitumor evaluation in several models to be able to reach a clear decision on potential for clinical development. Historically, this has been a major barrier to development of marine natural products, but the increasing emphasis on marine microorganisms as a culturable source of material is a promising development. Mariculture, directed biosynthesis, semisynthesis and total synthesis of the natural products or analogs can all be useful approaches. The isolation of the bryostatins from *Bugula neritina* was a difficult 20-year effort which is expected to bring bryostatin 1 to clinical trials. Progress was hampered by lack of a suitable sensitive and specific in vitro assay to monitor isolation. Bryostatin 1 is a powerful stimulator of hematopoiesis, which combined with its antitumor activity, could be extremely useful in chemotherapy.

The combination of more specific screening targets, more sensitive bioassays, better understanding of the chemical ecology of the organisms and improvements in culturing organisms and microbial associates provides hope that an increasing number of compounds for human trials in cancer will originate from the marine environment.

1 Introduction

The National Cancer Institute has had three major screening programs for both natural products and chemically defined compounds since the 1960s. These are shown below, with a few descriptive comments on each. Further amplification is given in the body of this chapter.

The first, operating from about 1958 to 1975 was developed as an in vivo screen (a compound-based system in the nomenclature of Boyd et al. [1]) utilizing the murine leukemia cell line, L1210.

The second, operating from 1975 to 1986 (also a compound-based system) was a variant of the first, but using an in vivo prescreen in the form of the murine leukemia cell line P388, followed by in vivo testing against an expanded panel of murine tumors and human xenografts in mice.

The third, operating from 1986 to-date, though modifications have and are being made on a continual basis (see discussion below and later in this chapter) is a disease-oriented screen [1] utilizing a panel of in vitro cell lines, predominantly human, but including the murine line P388 leukemia so that cross-comparisons can be made to prior data.

This screen is now undergoing further developmental work as a result of early [1986–87] experience, predominantly in the area of provenance and reproducibility of the cell lines involved, with the ultimate aim to produce an in vitro system whose component cell lines reflect the heterogeneity of human cancers, but whose lines are not "laboratory immortals".

Since the purpose of this paper is to present data on the methods of testing marine extracts and pure compounds isolated from marine environments, a discussion of the screening philosophies and the screens that were developed and used as a result of them, is not out of place. These systems with all their inherent flaws and biases, have been responsible for the identification and subsequent introduction into practice of very significant agents from all sources that are either in clinical practice or are in advanced testing (Phase I–III).

Thus, in this chapter, we will discuss the philosophies, show the screen develepements, discuss leads in progress and finally comment on the current and future testing and development of materials from marine environments.

2 Screening Methods

2.1 General Comments on Methods of Screening

Throughout the screening programs at NCI and also in the pharmaceutical industry worldwide, there has been, and still continues to be, a great debate amongst the concerned scientists and managers over the merits of a "defined mechanism-of-action screen" versus "general screening against suitable cell lines," both in vitro and in vivo. Both types of screens are inherently powerful. However, one must appreciate that with a mechanism-of-action screen, one can only find "actives" from whatever source, whose mechanism is directly related to the screen. The element of serendipity, to say nothing of what may be construed as "pharmacology", is missing.

This statement is not intended to be denigratory to the mechanism-of-action approach in any way. Screens for novel inhibitors of enzyme systems such as topoisomerases, specific sites of protein synthesis, interaction with expression of specific genes, etc., are inherently worthwhile, but if one is screening, or intending to screen, massive numbers of extracts and/or purified compounds, then ideally they should be screened against ALL enzyme/receptor systems which are key steps in the disease process. Hence the simple concept of an isolated enzyme/receptor screen may well become an unwieldy monster when assay numbers in the millions per year are considered, as happens to be the case when 5-15 different mechanistic assays are utilized per test sample.

These numbers are not unrealistic (see later) for a primary initial whole cell assay, but are compounded when isolated systems are considered because of the differing solvent/substrate controls required for such systems. On the positive side for isolated system assays, they are generally much faster than cell growth inhibition assays. However, in any screening system, the rate-limiting steps are generally those of sample preparation, and logging and interpretation of data rather than the intrinsic "speed" of the assay(s).

One primary consideration in the use of a whole-cell system is that "pharmacology", meaning general pharmacokinetic problems, is addressed, albeit partially. If the material does not cross the membrane, it will test negative. If metabolic activation is required to express activity of the compound, it has a chance of occurring.

2.2 Terminology and Inherent Assumptions

Prior to discussing the screens in detail, we should define both the terms and the assumptions inherent in them.

- Data from models must be relevant to the disease state in man:
  Such a statement seems obvious! However, every pharmaceutical house in the industry has discovered cures for diseases that are specific to the host animal used for the assay, and not to *Homo sapiens*.
- Antitumor: activity against a tumor in an animal model without undue toxicity towards the host.
  Such assays are always in vivo and a positive result implies that selectivity has occurred.
- Toxic/Cytotoxic: The former refers to harm to a living organism, the latter to inhibit or kill cells in an in vitro assay.
  In such assays, it is most important to distinguish between selective and nonselective (cyto)toxicity. Selective means that a given material shows significant differences in its toxicity towards cells of differing tumor types or between "normal" and tumor cells. Nonselective means either no differences or, as usually occurs, since only one cell type is used in the assay, no conclusions may be drawn.
- Screen: A device or mechanism for selecting compounds/extracts of interest from amongst a large population.
  This includes the following: "prescreen" (primary screen), in which a sensitive, high volume (usually in vitro) assay is used to select amongst very large numbers; "secondary" (detailed evaluation), in which selected agents are evaluated in depth using different models, dosing schedules, routes etc.; "bioactivity" monitor, a simple but not simplistic model that can be used to follow bioactivity during a physicochemical separation.

2.3 Applications, Characteristics and Strategies of Bioassays

2.3.1 General Comments

In Fig. 1 and Table 1 below we have given some of the key functions of bioassays. It is essential that one realizes that in any sequential system, the output of one screen is the direct input to another; thus any *faux pas* in the design and/or execution will

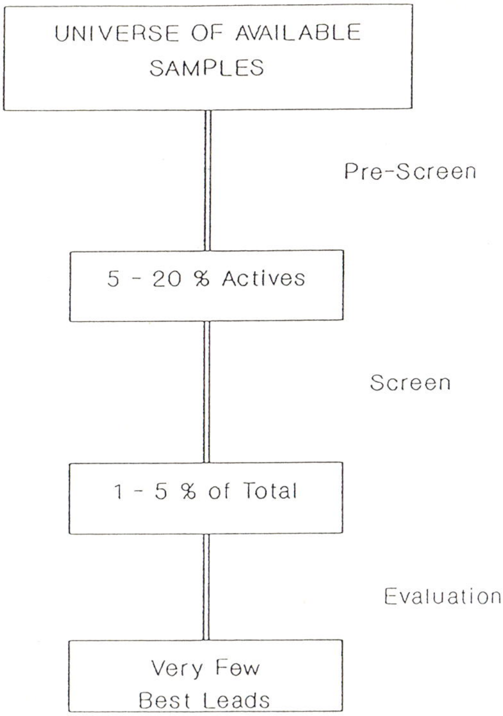

Fig. 1. Typical outline of screening process

"inversely multiply" in the next phase of the screen, leading ultimately to no active agents, irrespective of the numbers put through the system.

In any screen-mediated drug discovery and development program, as opposed to one of analog development, the overall objective is to start with a large universe of samples. For the purposes of this chapter, these are assumed to be extracts from marine organisms, both macro and micro, and pure compounds isolated therefrom; to distill from these, a few active and selective materials, which when purified and perhaps chemically modified, will show promise as useful agents for the treatment of cancer in man.

These selected and selective compounds will then undergo bulk isolation or synthesis, pharmaceutical development in the widest sense of the phrase, and toxicological studies before being readied for the first human trials.

Inspection of Table 1 and Fig. 1 indicates that the screening strategies using these types of screens are manifold and varied. Given a choice, different investigators will use the same "types" of screens but will vary the chronology, or establish different criteria depending upon their own bias(es) and/or experience. One must not forget however, that the sequence of screening should be logical and consistent with the hypothesis being tested. This latter point is significant, since the ultimate aim of any screening process must be that one can use the results to prove or

Table 1. Functions and characteristics of bioassays

	Prescreen	Screen	Secondary Testing	Isolation Monitor
Characteristics:				
Speed	Rapid	Modest	Slow	Rapid
Volume	High	Moderate/High	Low	High
Sensitivity	High	Moderate	Low	High
Complexity	Single protocol	Single Protocol	Specialized for Each Compound	Single protocol
Cost per Sample	Low	Low/Moderate	High	Low
Alternate Uses:				
	Monitor	Prescreen or Monitor	None	Prescreen

disprove a particular hypothesis. Unfortunately, too many screening programs tend to forget that essential point.

2.3.2 Hypotheses and Strategies

In the following section, we have given some examples of different hypotheses and the types of protocols/screens that we would consider using. We must emphasize again, that there is no "correct" way to screen; each group would modify to their taste; what we show below, are samples of how we might approach designing screening systems in response to different hypotheses.

Hypothesis A. A single sensitive in vivo screen will avoid the problem of too many in vitro positives that are in vivo negatives for whatever reason.

Prescreen: P388 leukemia, L1210 leukemia, Ehrlich ascites, or other single sensitive tumor as an in vivo model.

Screen: In vivo panel of more resistant tumor types.

Secondary: Additional testing against those tumors that respond to the earlier screen.

Monitor: P388 or other fast-growing cell line in vitro.

In this model, only materials that exhibit in vivo activity will be identified. The monitoring system, however, relies upon cytotoxicity; thus, if the in vivo activity is not due to cytotoxicity in the usual sense of the word, what is isolated may well be inactive in vivo.

Hypothesis B. Inhibition of mouse colon tumor cells in vitro will identify compounds that will inhibit human colon tumor cells.

Prescreen: Murine colon cell lines; measure growth inhibition, proceed with "actives".

Screen: In vivo in mice with either murine or human colon tumor lines, or both. Proceed with those compounds that significantly inhibit growth.

Secondary: Multiple cell lines, resistant/refractory lines, schedule of dosing, etc.

Monitor: Sensitive line from prescreen.

Hypothesis C. Compounds that selectively inhibit human colon tumor lines in vitro will inhibit these tumors in man. Obviously one may substitute any cell line source for "colon" without alteration of the hypothesis.

Prescreen: None, since this would require a large number of cell lines to indicate "selectivity"

Screen: Panel of human colon lines plus panels of tumor lines from other sites such as lung, breast, renal, etc., since function is to establish "selectivity".

Secondary: Same colon lines as xenografts in athymic mice, plus dose regimens.

Monitor: One or more of the sensitive or a sensitive and "non"-sensitive cell line in vitro to indicate that selectivity is retained.

In the above, the combination of resources normally allotted to pre- and regular screens permits establishment of selectivity at an early stage. There would be a higher probability that "actives" from such screens would have "selective" in vivo activity provided that pharmacokinetics permitted access of agent(s) to the tumor. Note also, that the secondary screening in this case is of "higher order" than usual.

Hypothesis D. Compounds which inhibit the action of topoisomerase in vitro should exhibit antitumor activity in vivo. (Nearly any biochemical assay could be substituted for "topoisomerase" without change in hypothesis).

Prescreen: Biochemical assay to measure enzyme inhibition.

Screen: Panel of mouse tumors and/or human tumor xenografts.

Secondary: Route, dosing schedule using "active" tumor lines; should also consider pharmacokinetics.

Monitor: As prescreen.

2.3.3 Assay Selection and Validation

In addition to proposing an hypothesis for assay selection and then designing the requisite experiments to prove or disprove it, there are other points that have to be borne in mind. At the risk of appearing flippant, we may codify these in the following set of rules.

Rule 1. Assays are always right; they are your most faithful friend and never lie. They always find what they are supposed to.

Rule 2. Assays are dangerous; they are literal and will lead one to severe embarrassment unless constantly monitored and questioned.

Rule 3. We can only validate, i.e., use positive and negative controls, assays that will pick up agents resembling those we know about.

Rule 4. The model closest to human disease will only see samples, compounds or crude extracts, that have previously been selected by less rigorous methods.

Let us now consider these points in detail, showing why they were made and how to live within their constraints.

Rule 1: Every group has experienced the frustrations involved when the "perfect compound" that has been selected on the basis of the assay turns out to be an artefact of the system used. With marine sources, macroorganism or microorganism, this is frequently the effect of inorganic salts upon the assay system, due in general to a lack of appreciation of the solubility of such compounds in the solvent systems used for isolation.

Rule 2: (Which in reality is rule 1 plus experience). What is usually forgotten during the excitement and frustrations of devising and running assay systems, is that one is working with a MODEL (surrogate, analog, imitation, or estimate) of the real issue . . . Disease in Man. In this respect, the essential difference(s) between the model and the intended user is(are) not always recognized early enough in the process. As an example of this, let us consider the use of a topoisomerase enzyme as a screen [Hypothesis D., Sect. 2.3.2].

If one uses a bacterial or lower eukaryotic enzyme as the source, any molecule found as result of the screen may well be inactive in the ultimate recipient . . . MAN, unless the sequences around the active site, any other binding domains and the mechanism of action and of inhibition resemble those in the human target cell. With the advent of genetic engineering and the ability to recognize, clone, and express mammalian enzymes, such problems will not be as difficult to overcome in the future as they have been in the past. However, at present, the information as to similarities and differences in enzymes from "normal" and tumor cells still has to be collected, collated and disseminated in order that the same problem does not continue to occur even though mammalian cells are being used.

Rule 3: A partial answer to the above problem, and others like them, is easily stated but it also has a major drawback. The answer at first blush is to validate the assay; i.e., drugs known to act at that site are used to show that the screen will recognize them under the assay conditions. The drawback is given in rule 3 above. Only drugs of known types will be picked up by a validated assay. For example in Hypothesis B of discovering highly selective drugs for colon cancer, there is no way to validate this in a positive sense because positive controls (drugs selective for treatment of colon cancer in man) do not exist. At best, one can carry out negative experiments by trying drugs that have undergone clinical trials for colon cancer in man and ascertain that none of them show strong selectivity in the model.

Rule 4: This is an example of the problems involved in deciding the biological level at which screening should be performed. Thus the fastest, cheapest and most target-specific high-volume assays are receptor binding or enzyme inhibition assays. In spite of these "advantages" these assays are also furthest from the human disease state. Models such as human tumor xenografts in athymic mice, though much closer to the human condition, are difficult, slow, expensive, of low capacity, and require significant prescreening in order to obtain a small enough number of samples to avoid exceeding the test capacity.

A very significant point in discussing the problems related to rule 4, and incidentally rules 2 and 3, is that in going from a mammal to an in vitro tissue or cell line and finally to an isolated enzyme/receptor system (*in practice one would proceed in opposite direction*), one loses what can be referred to as the physiological or pharmacological element in the equation. Thus, no account is taken in such a progression of the requirements for absorption, membrane transport, relative drug concentrations at site(s), metabolism, and excretion. Such parameters are the major reason why one loses initially active leads in proceeding from prescreen to secondary evaluation. We must also be aware that chemical derivatization can significantly alter the physicochemical parameters of a compound. Thus, novel compounds that prove inactive at a subsequent screening level should be re-evaluated as candidates for chemical manipulation.

2.4 Dereplication

Dereplication is the elimination of replicates or repeats of materials already isolated and known in the literature. Compounds that make it to the end of the "defined assay road" must have a suitable set of pharmacological parameters; for this reason, particularly in the case of natural products, the same or similar compounds continue to be rediscovered, thus leading to a very real requirement for dereplication at the earliest possible time in the sequence of operations.

Needless to say, no one wants to rediscover the wheel, but it is clear that many pharmacologically active substances have broad distribution in nature combined with high inherent activity and potency, which makes them easy to detect. It is fair to say that, unless one is screening an unusual group of organisms which have not previously been studied, or one is using a bioassay unrelated to any assay previously used for broad screening, one will rediscover known active compounds. Furthermore, it is a virtual certainty that if one screens common organisms by common types of screens, the dramatic leads will all be in known classes of compounds. For example, it is impossible to screen *Streptomyces* species by cytotoxicity-based assays without repeatedly finding actinomycins. Actinomycin D is a commercial antitumor agent which was discovered in the 1940s [2] and many derivatives and analogs are known [3]. Screening of microbial isolates is certainly not an efficient way to find a better actinomycin.

The challenge then is to find ways to dereplicate at an early stage, while the activity is still in the crude extract stage and minimal effort has been expended in isolation and structural elucidation. A full discussion of this topic is impossible in this chapter but for brevity, dereplication can be divided into three types:

Taxonomic dereplication – most secondary metabolites have a restricted distribution in nature and are characteristic of their source organisms at a particular taxonomic level (usually family, class or order for marine-derived metabolites). Conversely, if one knows the taxonomy of the active sample, one can readily find out what compounds, and especially what bioactive compounds, have been isolated from related organisms. For example, the likelihood that a cytotoxic agent from a sea cucumber (holothurian) is a steroidal saponin is quite high. One caution for marine organisms, however, is the possible presence of microbial associates which produce active metabolites. Such microbial products may not follow taxonomic relationships.

Biological dereplication – the bioactivity of the active material in an extract is characterized by its receptors. There may be many cytotoxic agents which occur in fish, but a sodium channel blocker in a puffer fish is highly likely to be tetrodotoxin. Patterns of bioactivity are more useful than single data points. An example taken from the dereplication program for discovery of antitumor antibiotics at the Warner Lambert – Parke Davis Company [4] is the use of a battery of 12–15 different test microorganisms including Gram-positive and Gram-negative bacteria, yeasts, and fungi. Zones of inhibition against all of these organisms are observed at a standard dilution of extract and the patterns of ratios of the zones are compared by computer with data from a large group of individual pure antitumor antibiotics in the computer memory using pattern matching programs. The best fit candidates are listed in order of probability. If there is no obvious match, the best possibilities can be further refined by chemical methods or by other bioassays. The main weakness of pattern matching becomes apparent, when there are two or more active agents of different types; in that case the patterns intermingle and do not give suitable matches. This problem will be resolved as the compounds separate during fractionation, provided the separated active fractions are dereplicated again.

Chemical dereplication – the chemical properties of a bioactive substance, which are detectable in mixtures, provide a basis for dereplication. These include UV-visible absorption maxima, color with spray reagents, mobility on chromatographic systems (tlc, hplc, or paper), acid-base properties, solubility, characteristic mass spectral fragmentations, and unusual NMR signals. The usefulness of these methods varies with the degree of crudeness of the sample and the chemical properties of the particular compound.

The total picture of dereplication is not a choice of taxonomic, biological or chemical methods, but rather is a combination of all of these, plus a good database, broad knowledge of the chemical and biological literature in natural products, a good memory, and intuition. In some areas of natural product research, such as the discovery of new antibiotics from soil microorganisms, rapid dereplication is so critical that no screening program could function without it. Among higher plants, where taxonomy is relatively well worked out, much dereplication depends on affiliations at the family, tribe, and even genus level. Until now, dereplication has not been a major issue in screening marine macroorganisms, because of the limited amount of literature data. Until fairly recently, every active compound isolated from a marine animal had a high probability of being new, but this is clearly no longer the case. Dereplication in the marine area may prove troublesome because of the difficulties in taxonomic identification of many biota, e.g. among the sponges, the paucity of established chemotaxonomic relationships, the lack of bioactivity data for known compounds, and the production of bioactive compounds by microbial symbionts and epiphytes. Access to a microbial dereplication data base is virtually essential. One international data base, whch the NCI helped to develop, is the Database of Bioactive Natural Products compiled by Janos Bérdy at the Institute for Drug Research in Budapest. It is searchable by producing organism, chemical class, UV maxima, color, molecular weight, and bioactivity or combinations of these parameters. It contains entries for over 14,000 compounds, mainly microbial products, although as of 1986 about 3,000 plant products and several hundred marine products are also included.

2.5 In vivo Models and Tumor Screens

These assays are expensive to perform, can take a relatively long time to complete, and because of the low levels activity generally found in natural products, are often not sensitive enough. Thus, a negative result in natural product screening has to be interpreted with care, particularly if it was positive in a prescreen. A recycling step, utilizing a concentrate of the extract, should be performed. Such a recycling system can be structured in a variety of ways; two possible protocols are shown in Fig. 2 below, where an in vitro prescreen is used as an example.

Note that in these protocols one performs one in vivo test in A versus two in B. One could drop the fractionation step in B, but if this is done, only those very active and/or abundant compounds would be (re)identified. The novel agent(s) would tend to be overlooked. This figure is a composite of the hypotheses previously described, and is deliberately left as such in order to demonstrate that a screening process is complex and not usually amenable to one simple scheme when one deals with natural products.

As is the case with in vitro models, there are certain parameters that should be understood when one discusses in vivo models.

Homograft: This model uses a tumor originating in the same species, and may be further subdivided if there are strain differences between the tumor line and the host. If the two strains are genetically compatible, the model is referred to as syngeneic; if different, allogeneic. Many of the latter type tumors will be rejected by the host's immune system, but after a longer period than the time for the tumor assay; such models are usually sensitive, particularly if the activity of the agent is partially or wholly mediated by the immune system. Allogeneic models may have artefactual problems; an example of a bioassay previously used at NCI is the Walker 256 carcinoma, which possesses high sensitivity to tannins. Hence, screening of plant extracts required tannin removal prior to the assay. Since this sensitivity was unknown before large scale use, a large number of actives had to be rescreened, following removal of polyphenols. Both syngeneic and allogeneic models [cf. Hypothesis A, Sect. 3.2.2] have been used in many research groups over the last twenty years; this large scale testing program was performed as a result of NCI contracts, for example. Xenograft: This model uses a host from one species and a tumor from another. Because immune systems recognize such tumor lines

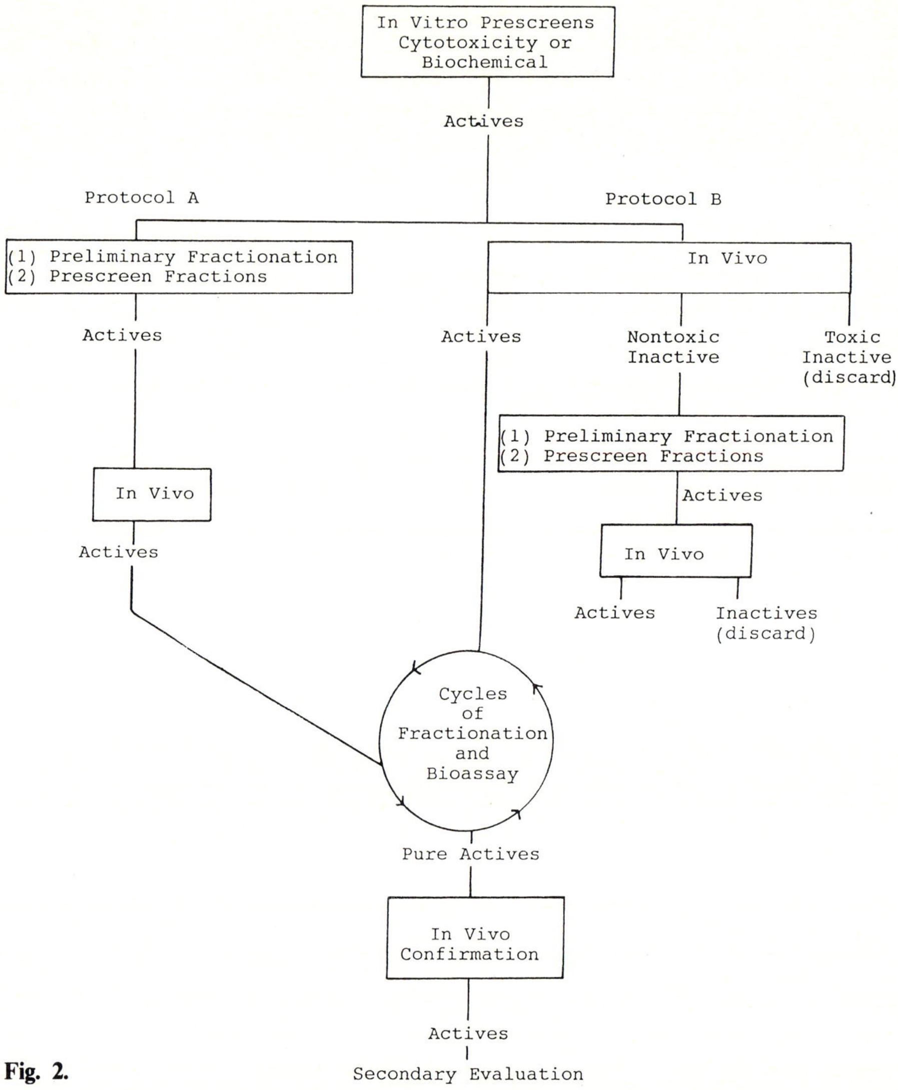

Fig. 2.

as non-self, in the past, rejection was a major problem and required the use of very high doses of immunosuppressive agents, which tended to obliterate any host-mediated response that aided the intrinsic activity of a given lead or extract. With the advent of athymic (nude) mice, testing against human xenografts became feasible. There are problems in such models, which are related to the lack of functional immune systems and the concomitant requirement for sterile technique at all stages, but these models provided a better chance of finding agents active against human cancer. To our knowledge, no research groups has advocated the use of a xenograft model as the primary screening system; it is in use, however, in the secondary evaluation portion of a system, as e.g., that which is currently used at the NCI.

Finally, in any in vivo model one must be cognizant of the significant differences in concentration of active materials found in an in vitro test versus the in vivo test of the same extract against the same or similar cell lines, since any active material must cross a multitude of barriers before reaching the site of action. The huge dilution factor that is involved when working with the volume of a mouse versus the volume of a well in a cell culture dish must not be neglected either.

3 Screening at the National Cancer Institute

3.1 Results from the NCI Screens Prior to 1986

As mentioned briefly in the Introduction, the NCI for the past 25 years has either performed, contracted or directly funded, the testing of a vast number of compounds in the process of looking for agents that are useful in the treatment of human cancer. Agents tested have run the gamut from pure synthetic compounds to natural product extracts or compounds from animal and plant sources, including fermentation products of terrestrial and some marine microbes as well as marine macroorganisms.

The screening philosophy used during the major part of these studies before 1986 can be best described as compound oriented. This aim can be further stated as the discovery of:

> cytotoxic agents, which destroy rapidly growing tumor cells, and hence may be clinically efficacious against human cancers.

This statement actually contains a significant amount of circular logic, since the agents that were found as a result of these studies and were clinically useful were active mainly against rapidly growing cancers such as leukemias, irrespective of the chemical class and mechanism of action.

Initially, the screen used was of the type described earlier in hypothesis A; L1210 leukemia was the tumor line that was used. Inspection of the results of this screen showed that although more than 50% of agents ultimately selected for clinical study that had been detected initially by this system showed positive activity in clinical trials [5, 6], the types of tumor inhibited were only those that grew very rapidly; the leukemias, lymphomas, a few discrete cases of choriocarcinoma and Ewing's sarcoma [7]. There was essentially no activity against slow growing solid tumors, particularly lung, colon and mammary cancers which are the tumors with the highest morbidity and mortality in man.

An alternative strategy was then proposed for a variety of reasons in addition to those stated above. These included the advent of the athymic mouse models, the requirement to screen even larger numbers of compounds [10,000 per year] and the increased cost of in vivo screening due to the increase in the number of in vivo models projected for studies with the more refractory tumor lines. Thus the

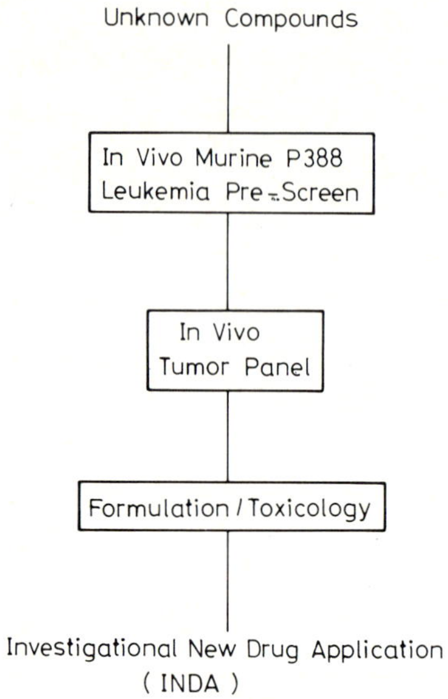

Fig. 3.

strategy was revised as shown in Fig. 3 to utilize an in vivo prescreen followed by more rigorous in vivo models.

Inspection of the murine cell lines available at that time indicated that the P388 murine leukemia line was sensitive to most of the clinical agents then in use, with exception of bleomycins and steroids. In addition, this line was also more sensitive than L1210 and could be established as an in vivo model in a variety of sites in the animal, thus permitting "pharmacology of the agents" to be tested under varied conditions.

In order to reduce the bias that is inherent in the use of a single cell line at all stages of the testing, the following steps were taken. An initial check which was followed throughout the use of the scheme, indicated that the P388 line was not obviously discriminatory, though the materials ultimately identified were still active only against fast growing tumors. A two-stage process was also used in the evaluation of data; only a 20% increase in lifespan was required for initial activity, and any compound that had a known antiproliferative or a biochemical mode of action that was of interest was admitted to the secondary test system irrespective of its activity in the P388 model.

By use of this system, the pool of 10,000 compounds per year was whittled down to a much more manageable level of approximately 500. Of these, approximately 200 entered the next stage of the process, the expanded panel of initially murine tumors followed by the use of the subrenal capsule technique with human tumor xenografts in athymic mice. In these expanded evaluations, two levels of activity were recognized; the first was an extension of lifespan to 125–140% or a reduction in tumor weight to <40%, both with respect to controls; variation in activity criterion was dependent on the cell line. The second level was an increase in lifespan to $>149\%$ or tumor weight reduction to $<20\%$. Only those compounds or extracts that showed the second activity level were considered for further scrutiny.

The lines used in the prescreen and the secondary tumor panel are shown in Tables 2 and 3 below.

We should point out at this stage that, in addition to looking for active materials, NCI was also using the secondary tumor panel to partially answer more fundamental questions related to screening strategies.

Thus one goal was to measure the productivity of a screen by using the number of clinically active agents as the marker, and by checking if there was any correlation between murine and human tumors that were nominally derived from the same site. (An extension of Hypothesis B, presented earlier.)

Although the full analysis of the results from these studies is not available, since some of the drug entities identified are still undergoing clinical evaluation, Venditti et al. [8] were able to draw the following preliminary conclusions.

1. The order of sensitivity towards tumors in models paralleled clinical experience with the drugs: the number of actives was in the order leukemia > melanoma > mammary > lung > colon, with very few active compounds against murine or human lung or colon tumors.
2. There was little cross-species site correlation in activity. Compounds found to be active at one site, would tend to be active in all sites in a given species; but activities in the other species would vary from none to all.
3. Compounds that exhibited activity primarily against the human mammary xenografts were not active against the murine tumors.

Although, as mentioned above, a full analysis is not yet possible, it has become obvious that the outcome will be inherently biased as a result of using one murine

Table 2. NCI prescreen and tumor panel models

Tumor	Test Host	Inoculum	Site
Mouse Tumors			
B16 melanoma	BDF_1, $B6C3F_1$	1:10 brei	ip
$CD8F_1$ mammary carcinoma[a]	$CD8F_1$	1:20 brei	sc
Colon 38 carcinoma	BDF_1, $B6C3F_1$	70 mg fragment	sc
L1210 leukemia	BDF_1, CDF_1	10^5 ascites cells	ip
P388 leukemia	BDF_1, CDF_1	10^6 ascites cells	ip
Lewis lung carcinoma	BDF_1, $B6C3D_1$	10^6 cells	iv
Human tumor xenografts			
CX-1 colon	nu/nu Swiss	$1 \times 1 \times 1$ mm fragment	src[b]
LX-1 lung	nu/nu Swiss	$1 \times 1 \times 1$ mm fragment	src[b]
MX-1 mammary	nu/nu Swiss	$1 \times 1 \times 1$ mm fragment	src[b]

[a] $CD8F_1$ mammary tumor tests are conducted in mice bearing the first transplant generation from the spontaneous tumor.

[b] Subrenal capsule

Table 3. Tumor panel models: treatment and evaluation factors

Model	Drug Route/ Schedule	Parameter	Active T/C (%)[a] (+)	(++)	Evaluation Day
		Tumor Panel			
Mouse tumors					
ip B16 melanoma	ip/Q1D, D1-9	Median survival time	≥125	≥150	60
sc $CD8F_1$ mammary	ip/Q1D, once[b]	Median tumor weight change	≤20	≤0	30
sc Colon 38	ip/Q7D, D2-9	Median tumor weight change	≤42	≤10	20
iv Lewis lung	ip/Q1D, D1-9	Median survival time	≥140	≥150	60
ip L1210 leukemia	ip/Q1D, D1-9	Median survival time	≥125	≥150	30
Human tumor xenografts					
src CX-1 colon	sc/Q4D, D1-13	Mean tumor weight change	≤20	≤10	15
src LX-1 lung	sc/Q4D, D1-9	Mean tumor weight change	≤20	≤10	11
src MX-1 mammary	sc/Q4D, D1-9	Mean tumor weight change	≤20	≤10	11

[a] For survival assays, T/C is computed as survival time of treated mice × 100 ÷ survival time of controls. For tumor inhibition assays, T/C is computed as change in weight of treated tumor × 100 ÷ change in weight of control tumors.
[b] Treatment given when tumors reach weight of 100–700 mg.

tumor line, a sensitive, fast growing, leukemia, as a prescreen and one or two cell lines as the exemplar of a given tumor class in the follow-up testing.

Perhaps the latter bias is the most significant, since it was shown by many workers [see references in Boyd et al. [1], that solid human tumors from a given site are composed of a heterogeneous population of cells. The use of one to three individual lines cannot be considered a valid representation of the clinical state, and thus does not represent a particularly valid screening philosophy. Too many in vivo models would be required to produce data that would have a reduced or minimized bias; hence the screening philosophy had to be revised to take into account the findings from the previous 20 years of work.

3.2 Disease-Oriented Screening Systems

Instead of the compound orientation of the previous system(s), the screening protocols instituted at the NCI in 1986 were designed to take into account the heterogeneity of the individual clinical entities and perhaps can best be described as a disease-oriented screen. The clinical rationale behind such a change is not a suitable subject for this chapter, but it has been described in depth in the case of lung tumors and subsequently been extended to other types [1, 9, 10]. Suffice it to say, the screen was designed for rapid in vitro assessment of materials against multiple sensitive and refractory lines. The test protocols included both established lines from collections and initially primary growths from clinical specimens, with the overall aim of identifying agents with SELECTIVITY towards a particular tumor type, when compared to its activity against other tumor types. Such agents would then be tested in vivo against the same or similar tumors as xenografts in athymic mice, followed by expanded testing [cf. Hypothesis C, in Sect. 2.3.2]. Implicit in this scheme was the decision that any compounds found to be selective would be placed on a fast-track towards clinical testing. This would include setting up systems to produce large quantities of material of defined purity, early toxicology, preparation of chemical analogs if necessary and screening for other indications, as supplies became available.

The overall schematic of the first phase of the process is shown in Fig. 4.

Although the concept is relatively simple, the logistics of the operation are significantly difficult, because of the following considerations.

The desire to screen in excess of 10,000 extracts and pure components per year over a 5- to 6-fold dose range in duplicate or more would mean that one cell line would generate approximately 100,000 assays, not counting controls.

The desire to utilize 60 to 100 cell lines of varying type would expand this number to approximately 10^7 assays per year.

Since in vivo assays are impossible, in vitro assays require a simple reproducible test for cell death or growth inhibition. In other words, it remains a cytotoxicity assay.

Cell lines to be used must be thoroughly characterized and free of adventitious agents.

The assays must be simple, capable to be automated and the data handling must be computerized.

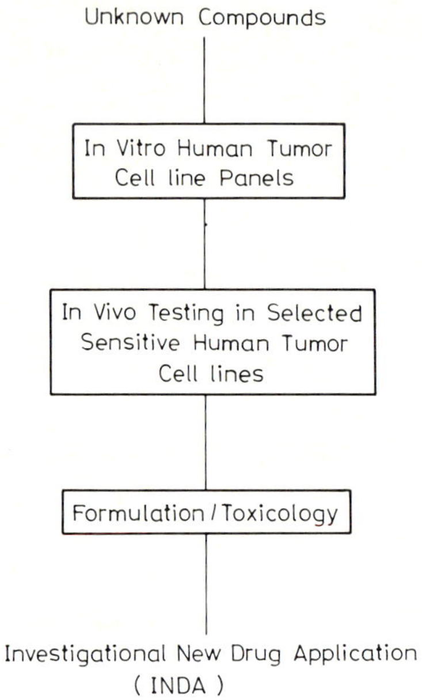

Fig. 4. In vitro – based disease-oriented screen 1986

An assay system was developed, that took into account the logistical problems outlined above. The basis was a modification of the colorimetric assay originally described by Mosmann [11]. This assay is based on the ability of a respiring cell to reduce a tetrazolium salt [MTT] to a formazan dyestuff. Coupling this biochemical response to 96-well plates and automated "ELISA" plate readers, permitted the establishment of a simple and reproducible assay for cytotoxicity, which had the potential to screen very large numbers of samples for selective toxicity [12].

There have been a variety of problems in developing this type of assay, a detailed discussion of which is beyond the scope of this chapter. In brief, the problems are: growth of many diverse cell lines varies in the same culture media; diverse cell lines have inherently different capabilities to metabolize the dye reagent; well characterized human cell lines with known histories are difficult to find for many types of tumors; with 50–60 cells lines in one screening program contamination is always a threat; and the huge volumes of data generated require extensive computer programming.

Despite all these difficulties, screening at a rate of 5000 samples per year in 50 cell lines began in Summer, 1988, with anticipated expansion to 10,000 samples per year in 100 cell lines in the future.

4 Marine-Derived Antitumor Compounds

Although the chemical literature is replete with descriptions of novel agents isolated from a multiplicity of marine macro- and microorganisms [13, 14], the description of their biological activities leaves much to be desired. Perhaps the major reason for this is that in most cases, the agents were isolated as novel chemicals that proved to be new and novel structural types rather than from extracts with biological activity.

Only a handful of commercial enterprises have looked at the marine environment as a source of novel pharmacological agents. These have included the Roche Institute for Marine Pharmacology in Australia, Wellcome in the UK and USA, CIBA-Geigy in Europe, and SeaPharm in the USA. None of these groups are now active, though all used biologically guided isolation systems.

There are very active academic groups who use some form of biologically monitored isolation processes to obtain novel compounds from marine organisms. In general, these can be described as compound-based in the sense that they are looking for activities using a prescreen with P388 or another sensitive cell line in vitro/in vivo and follow activities by in vitro activity against P388. These include the University of Hawaii (cyanobacteria), University of Rhode Island (algae), Scripps Institution of Oceanography (macro- and microorganisms), the University of Canterbury, New Zealand (macroorganisms), and a group at the Institute of Microbial Chemistry in Tokyo, (microorganisms). Significant novel structures with antitumor activities in murine models have recently been reported from these groups, e.g., the mycalamides, which were described at the 1988 American Association for Cancer Research meeting [15]; they were isolated and purified by the New Zealand group from a sponge of the genus *Mycale* by following bioactivity against P388 [16].

Active agents that were identified as a result of NCI screening of over 18,000 extracts of marine organisms in the last 20 years using the in vivo P388 model, include the didemnins, isolated from the tunicate *Trididemnum* sp. [17], one of which, didemnin B (see Fig. 5) is currently in clinical trials [18]. This material also has antiviral and immunomodulating activities. It has also recently been synthesized [19], thus opening up the potential for chemical modification if desired.

In addition to this compound, bioassay-directed fractionation of a sea hare extract, *Dolabella auricularia*, resulted in the cyclic peptide family known collectively as the dolastatins [20]. The structure of one of the most active antitumor constituents, dolastatin 10, has recently been elucidated [21]. The possibility that these metabolites are in fact derived from an ingested alga should not be overlooked, since the metabolites obtained from other sea hares are almost exclusively derived from such sources [13]. In addition to the above, active research and development at NCI is now underway on large-scale extraction and isolation of the macrocyclic polyether lactones known as the bryostatins. These are found in very low yield ($<10^{-6}\%$) [22] in the bryozoan *Bugula* sp. and have potent activity in the P388 model and strongly affect protein kinase C [23]. Their structures were determined by both X-ray and complex NMR experiments [22, 24–26]. The structure of bryostatin 1 is shown in Fig. 6. A more detailed discussion is found in Sect. 6 of this chapter.

In addition to these compounds from macroorganisms (some could be metabolites from ingested microorganism), there are other novel materials isolated from microorganisms that have been identified as active in antitumor screens.

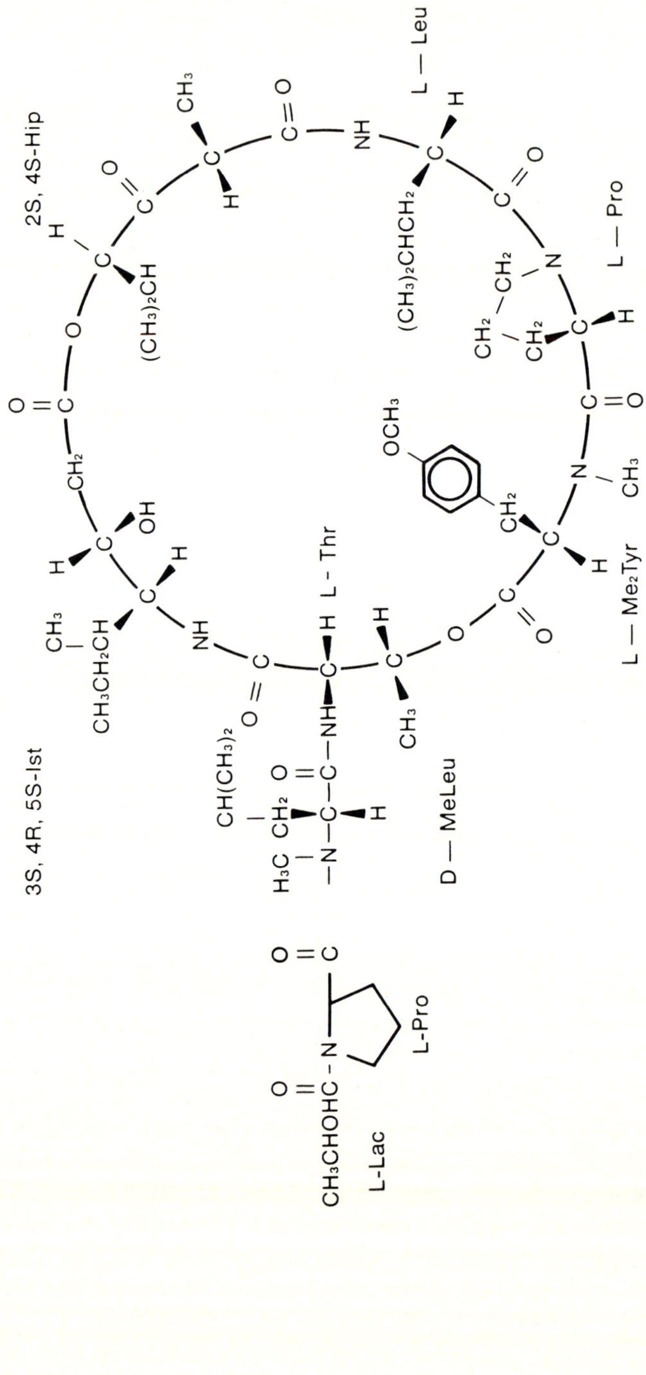

Fig. 5. Didemnin B

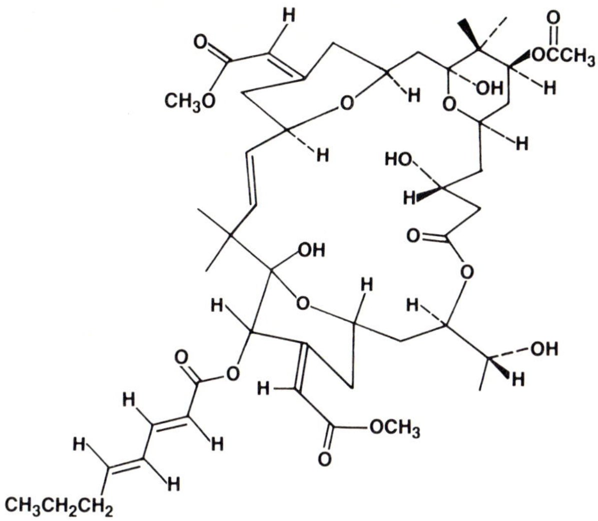

Fig. 6. Bryostatin 1

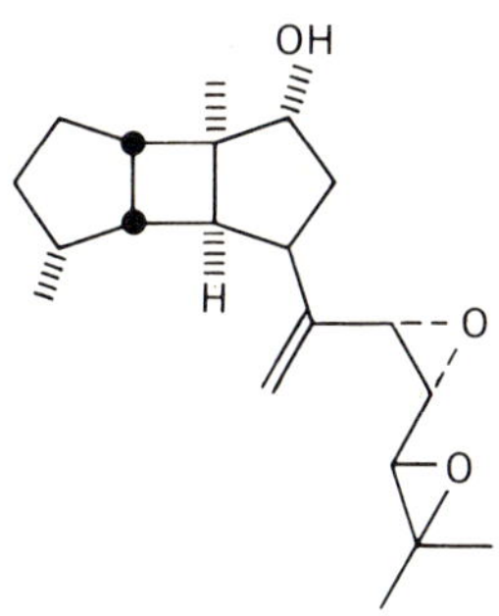

Fig. 7. Spatol

One example is the diterpene spatol (Fig. 7) [27] from the brown alga *Spatoglossum schmittii*, where the potent cytotoxicity may well be due to the 1,3-diepoxide moiety in the sidechain.

Others are elaborated by blue-green algae (cyanobacteria). These organisms are now being utilized as sources of novel potent cytotoxic agents, with a large amount of work being reported from the group at the University of Hawaii [28]. Work from this group has shown that the antineoplastic activity of *Lyngbya majuscula* is due to debromoaplysiatoxin [29] and it, together with a related compound, oscillatoxin A, are the active principles elaborated by *Schizothrix calcicola* and *Oscillatoria nigroviridis* when grown as a mixed culture [30].

5 Current and Future Directions in Screening Marine Natural Products

As we have shown, all of the current crop of active materials now undergoing advanced testing, have come from screens that, although occasionally modified, were in essence compound-oriented with murine tumor lines as the basis of the detection systems.

With the recognition that such systems have tended to produce only materials active against fast growing cell lines, future testing at the NCI will concentrate on in vitro disease-based systems described earlier in this chapter. Only materials that demonstrate significant specificity for a given class or cell line will be further investigated for their ultimate potential in humans.

Neither the screens nor sources of marine-based extracts will remain static. There will be feedback into the screen design on the basis of the types of chemical compounds identified as time progresses. As mentioned in an earlier section, there are no current agents identified to validate this approach; it is the aim of the program to find such materials.

In addition to retrospective evaluation based on the chemical nature of actives, the types of cell lines and their sources will be scrutinized for clinical relevance using both in-house and outside experts in the field. All lines will have full characterization before use, since we have no desire to find a cure for a tumor that is a laboratory artefact.

With respect to sources of extracts and compounds, the NCI will continue to solicit samples, both purified and crude extracts from all sources; in addition, by the contracting process, it will obtain a broad representation of samples from as wide a geographic area as is feasible.

In order to expand the sources, NCI has funded under the Small Business Innovative Research program phase I studies in macroorganisms from the intertidal areas, in marine microbes as a source of antineoplastic agents; it has established a long term contract to isolate, grow and test extracts from large numbers of Cyanobacteria, and will shortly award a contract to grow marine Protozoa. Contracts have also been awarded to scale up production of organisms of interest.

Contracts for collection of up to 5,000 different marine species from deep water and a similar number from shallow water in the Indo-Pacific were established in 1987. The deep water contract is currently halted until the samples collected to date have been evaluated for activity, but the shallow water collection program is continuing.

The aim of all these NCI-funded collections and of others that will be directed towards different types and/or sources of marine organisms in the future, is to tap into the vast, unexplored potential of marine biochemistry for novel bioactive leads and to bring the best compounds into development.

Finally, the necessary infrastructure has been established at the Frederick, MD, Cancer Research Facility to deal with the requirements for storage, extraction, and testing of the submitted samples, and to provide the resources required for work up of the large volumes of macroorganisms and microorganisms that would be required to isolate sufficient material from an active extract for clinical testing.

6 Development of Active Compounds

6.1 Issues of Long Term Bulk Supply

The road from discovery of an active compound to development presents a huge barrier which few compounds, either natural or synthetic, are able to cross. In the case of marine products, the situation is even more difficult than usual. Before any pharmaceutical company will consider placing a product into development it will want assurance that there is nearly 100% certainty of availability of large scale supply for marketing the drug, not merely a sufficient supply for preclinical development or preliminary human trials. The logic is inescapable; why invest the time and expense of development, if there will not be adequate drug supply for marketing later on? What is desired, is a supply that is totally controllable by the company; on this basis we can divide marine-derived drug candidates into two groups, those which are fermentable and those which are not.

For compounds produced by nonphototrophic microorganisms, possession of a productive culture solves the problem; supply can be produced in large fermentors at will (at least in principle). There are, however, many difficulties in the fermentation of marine microorganisms, including their tendency to grow only to low cell densities, thereby producing little mass of either organism or metabolites; their need for high salt concentrations which are corrosive to equipment; fairly low temperatures for optimum growth, which result in very long fermentations that tie up equipment and increase the risk of contamination; and the lack of knowledge, which fermentation parameters are most important for various types of organisms. However, many pharmaceutical companies are currently undertaking microbial fermentations and are willingly accepting the difficulties in return for the prospect of controlled supply.

In other groups of marine macroorganisms the supply problems are totally different. First, we should consider the possibilities for obtaining bulk supply sufficient for marketing; total synthesis, semisynthesis from readily available precursors and synthesis of simpler analogs are possible, but in the case of a complex chemical structure with a substantial number of asymmetric centers the possibilities are reduced to collection and cultivation. Neither of these are desirable prospects; collection from natural sources generally does not secure the supply of any drug and, with minor exceptions for low volume and hence not very profitable drugs, cultivation must be undertaken. By analogy, in the case of drugs from higher terrestrial plants, nearly every drug plant is cultivated, whether it is a tree, such as *Cinchona* sp., for quinine production, an herbaceous plant, like *Catharanthus* sp., for vincristine or a bulb, like *Colchicum* sp. for colchicine production. Most American pharmaceutical companies view reliance on a cultivated plant with some trepidation unless it can be cultivated in the United States or plantations can be set up readily in a number of countries in different parts of the world. The fear is that reliance on a single foreign source can become a disaster in the event of war, change of government, embargo, plant disease, insect infestation, unreasonable export duties, etc. In such a situation a company is faced with the unpleasant choices of abandoning the product, providing it at a loss to patients who continue to need it, or raising prices dramatically and losing market share to other products with similar therapeutic properties. Added to these basic issues of cultivation are questions, whether many marine organisms can even be grown in culture situations, what the costs are apt to be, whether mariculture of many groups of organisms is scalable to production quantities, what the roles of diet and the food chain are, how changes in diet with season or water conditions affect production, the role of microbial associates in production of the desired metabolite, etc.

Based on the above considerations it is not at all surprising that the macroorganisms from marine environments have not been regarded very seriously by most companies as direct sources of drugs for the marketplace. There is increased interest in marine animals as sources of new lead compounds, however. The benefits of this interest to the marine research community are as yet unclear.

Until the 1980's little drug discovery work on marine microorganisms was undertaken with the exception of pioneers such as Okami at the Institute of Microbial Chemistry in Tokyo [31]. The interest of the pharmaceutical industry is moving in that direction, as is the interest of many academic researchers. Impetus for microbial research has come from discoveries that key bioactive metabolites such as saxitoxin [32–33], okadaic acid [34, 35], tetrodotoxin [36], and others are products not of the macroorganisms where they were originally found, but rather of the associated microbial flora.

6.2 Supply of Compound for Biological Testing

Over the past twenty years or so, we have seen a profound change in the way natural products chemistry is being carried out. In the 1960s we often needed 30 to 40 mg of a compound for a decent proton NMR spectrum and carbon spectra were obtainable only with hundreds of milligrams of sample. With the tremendous advances in both spectroscopy and separation science over the last two decades, it has become routine to perform complete structure elucidations on a few milligrams of sample or even less. While these advances have permitted much excellent chemistry, which was previously unattainable, there has been one major disadvantage for drug discovery – the paucity of isolated sample. Developments in drug screening methodology have kept pace with chemistry and separation science and we now have exquisitely sensitive receptor binding assays that can detect nanograms of an active substance and can run at high sample volumes. The size of the mouse has not changed, though, and it takes about the same amount of material to do an in vivo assay as it did in 1960. This frequently leads to an impasse, where in vitro activity is present in an interesting molecule and there is no follow-up. This commonly occurs when an academic investigator discovers a compound of novel structural type and has a small quantity available for screening by pharmaceutical companies. In vitro activity is found, but the original investigator does not have the desire or capability to scale up the original isolation 20–50-fold and the company does not want to use its own personnel or invest funds and time on an in vitro lead, which could well be inactive in animal models. Even if material can be scraped together for one test in vivo, the result is often not definitive. A negative test in one model is not conclusive and often, if the first in vivo test is positive, it may only be weakly or moderately positive, because there was no advance knowledge of dosing quantity or frequency. There are no easy answers to this problem of compound supply for the university chemist. The NCI has taken the position of making large enough collections (generally one kg of organism) to afford substantial amounts of extract and to follow up all promising leads by purifying additional amounts of extract.

6.3 Scale-up Isolation – Bryostatins

Very few marine products have been scaled up for development as drug candidates and we wish to conclude this chapter by discussing one marine product of high current interest in order to give the reader a feeling for what is involved in such a project.

6.3.1 Lead Discovery and Isolation

One of the most interesting compounds from any source, natural or synthetic, now in development at the NCI is bryostatin 1. The biological data for this compound have continued to become more and more interesting, as its effects at the molecular level on various biological systems have been examined. Before a discussion of the present status of the bryostatins, some background is in order. Pettit et al. [22] in their paper on the isolation and structure of bryostatin 1, state that this work was the culmination of a project started in 1968. We wish to discuss first the events that occurred between 1968 and 1982 from the NCI standpoint, then the data and events that led to the decision to develop bryostatin 1 as a clinical candidate in the U.S., and finally the current status of the project.

The original collection of the bryozan *Bugula neritina*, which was found active, was made by Jack Rudloe of the Gulf Specimen Company, off the west coast of Florida in June, 1968, as part of a contract with the NCI to collect a variety of marine specimens for screening. The sample was shipped to Professor Pettit at Arizona State University who prepared the aqueous isopropanol extract that was submitted to the NCI for screening. This sample, NCI screening number B705008, was tested against two in vivo models, L1210 leukemia and P388 leukemia. The extract was inactive and nontoxic to mice infected with L1210 leukemia at doses of 400 mg/kg/day, as tested in April and July of 1969, on two schedules (single injection and daily for nine days). In May of 1969, an experiment on mice infected with P388 leukemia showed a 68% increase in life span (ILS) in the treated group at a dose of 400 mg/kg/day for 10 days. A duplicate test was begun immediately and at that time an ILS of 100% was seen, although the experimental controls were not fully satisfactory. Two subsequent dose-response experiments against P388 leukemia were conducted in July and September of 1969, during which doses ranging from 800 mg/kg/day to 100 mg/kg/day were administered for 10 days. These showed ILS values of 50% to 90% at all doses tested, thus demonstrating that the active material in *Bugula neritina* had activity over a broad dose range. Furthermore, even at 800 mg/kg/day there was little toxicity, indicating the possibilities that the active component might be broadly active and nontoxic; abundant and not very potent; or potent but present in very small concentrations. The original extract was also tested for cytotoxicity towards KB cells and was non-cytotoxic at the highest concentration tested, 100 μg/mL; this was potentially troublesome, since the purpose of this assay was to find an in vitro system in order to monitor fractionation of the active extract. Lack of cytotoxicity would mean that all fractions would have to be tested in vivo, a very slow and expensive process.

Since the original collection was not of sufficient size to permit fractionation, especially as about 8–10 g of extract had already been consumed for in vivo assays, a recollection was made in the Spring of 1970 under NCI sponsorship. This recollection and a number of partition fractions derived from it were tested in July of 1970 and found to be completely inactive against P388 leukemia and against a solid tumor model, the Walker 256 carcinosarcoma. Another collection of *B. neritina* made in 1971 showed modest ILS values of about 45% against P388 leukemia, but individual fractions were inactive. The impression at that time (mid-1972) was that there was some activity which might be unstable and which was not very reproducible from one collection to another. The project was put aside in favor of more promising leads.

The issue of reproducibility of activity in recollections is a troublesome one, which is by no means unique to the marine environment. It is also of significance in working with terrestrial plant collections and in regrowing or scaling up microbial cultures. In the broadest sense of the word we can attribute much of this variation to the ecology of the organism and the seemingly fundamental but difficultly demonstrated principle that organisms do not waste energy making compounds which they do not need. Thus production of many biologically active secondary metabolites can be hypothesized to occur in response to various stressors in the environment, either of a physical or chemical/biochemical nature. Physical stressors for a marine organism could include tactile stress, changes in water temperature, currents, or light levels, while chemical stressors would include toxins and various kinds of inhibitors of biological processes released by competitors. Other key factors are the nutritional state and energy balance of the producing organism and its position in the food chain. Another unresolved question is the possible role of microbial associates as producers, transformers, or sources of precursors for chemical stressors. There are obvious similarities between certain sponge metabolites such as okadaic acid [37] or acanthafolicin [38] and polyether antibiotics of microbial origin, and one must ask what is produced by the macroorganism, what by the microbial associate, and what by the combination of the two. Okadaic acid e.g. was later isolated from a microorganism [34, 35]. To return specifically to *B. neritina*, a fair amount of literature exists on growth conditions and life history [39–45], but to date there is no known correlation between growth, living conditions, and concentration of bryostatins. NCI is currently sponsoring limited field work to collect data on habitat and seasonality versus bryostatin content in *Bugula*. We can make educated guesses using production of antibiotics by microorganisms as a model. For a great majority of antibiotics isolated from *Streptomycetes*, the microbes, with a full complement of nutrients and plenty of space (modest inoculum level), grow in log phase and produce very little antibiotic. Antibiotic production generally occurs at later stages, where nutrients and space or air exchange begins to be limiting, or when the growth causes changes in the medium such as pH shifts. Furthermore, many microbes can produce several different and chemically unrelated antibiotics, depending on growth conditions and presumably the exact nature of the stressors; can we expect marine macroorganisms to behave likewise?

The next chapter in the bryostatin story began in March, 1976, when Pettit and a group of his coworkers made a series of collections of *B.neritina* in the Gulf of California in the State of Sonora, Mexico. This sample was screened by the NCI against the P388 leukemia in January, 1977, and was found active (ILS 67%) at a dose of 100 mg/kg/day on a daily schedule for 9 days. Doses of 200 and 400 mg/kg/day were toxic. Confirmatory testing shortly thereafter showed good activity (ILS range 50–100%) over a range of doses from 44 to 150 mg/kg/day. This same extract was not cytotoxic to KB cells. Again, there was not enough material available to begin fractionation in a serious way, and work was postponed until a recollection of the Sonoran material could be obtained. Recollection in 1978 and 1979 was not possible due to damage to the habitat by severe storms; in late 1979 it was decided to go back again to the old Florida recollection. When all fractions were recombined and repartitioned, modest activity was observed in the P388 in vivo assay. This group of fractions was also tested in the P388 cell line for cytotoxicity and cytotoxicity was observed for the first time. It appeared that the P388 cell line was much more sensitive to the *B. neritina* compounds than the KB cell line. This possibility was interesting, as one of us had been examining numerous data from NCI screening of plant extracts and had noted that a group of plants in the families Euphorbiaceae and Thymelaeaceae were disproportionately active in the P388 in vivo screen, and that they were quite cytotoxic to P388 cells, but not KB or L1210 cells in culture (NCI, unpublished observations). Several of these plants were subsequently recollected and fractionated by Farnsworth and Cordell's group at the University of Illinois, Chicago and by Pettit's group at Arizona State University, yielding various phorbols and daphnanes. The possibility of a phorbol-like structure from *Bugula* seemed remote, however, since no compounds of either phorbol or daphnane types had ever been isolated from marine organisms.

Meanwhile, a new collection was made in Sonora by Pettit and coworkers which was tested at the NCI in March of 1980. This material was quite potent, being totally toxic in vivo at 50 mg/kg/day in the P388 system and subsequent fractions were active at doses as low as 3 mg/kg/day. Further fractionation guided by P388 bioassays both in vivo and in vitro led to isolation of bryostatins 1 and 2 in January, 1981; subsequently Prettit et al. isolated a large number of bryostatins from both Florida and Gulf of California collections [22, 24–26, 46–49]. There was some difficulty in following the in vitro activity. Fractions which were most potent and had greatest antitumor activity in vivo were less cytotoxic than neighboring fractions that were less potent and less active in vivo. Thus in vivo testing had to be used to make key decisions during fractionation. Initially, two possibilities were envisaged. Either there were problems with the assay, or closely related cytotoxic compounds were devoid of in vivo activity. The assays, however, were reproducible and, strangely as more bryostatins were isolated, they almost always were markedly less cytotoxic to P388 cells when tested as the pure compounds than when tested as impure fractions. The possibility of solubility problems was considered, but this could not be responsible for the large differences observed. Very recent results from Pettit's and Blumberg's laboratories [50] showed that dilution of pure bryostatins, first increases cytotoxicity to P388 cells and then gradually decreases with further dilution. Apparently the receptors become saturated at high concentrations and the on–off binding and release necessary for response are lost. While saturation of receptors is a well-known phenomenon *per se*, we had not observed it previously in a cytotoxicity assay with mammalian cell lines. Generally in such assays the cell killing event is proportional to binding to the key receptor, which in turn is proportional to drug concentration. To express it differently, no saturation is observed, because the initial binding leads to an irreversible cell damaging event.

The bryostatins have also been isolated from three other organisms, the bryozoan *Amathia convoluta* [51], the sponge *Lissodendoryx isodictyalis* [52], and the tunicate *Aplidium californicum* [53]. These are important examples for the lessons which they provide. *Amathia convoluta* is a bryozoan that was collected in the Gulf of Mexico from similar locations as *B. neritina*. Again, there was a history of erratic activity in recollections but Pettit et al. succeeded in isolating bryostatins 4, 5, 6, and 8 in small amounts from a 1981 recollection [50]. Analysis of voucher samples of various collections showed that 0.2% to 3% of the dry weight of the animal in fact was *B. neritina* which grew as an epiphyte on *A. convoluta*. While the concentrations of bryostatins 4, 5, and 6 could be accounted for by the presence of *B. neritina* in the collection, the concentration of bryostatin 8 was disproportionately high and Pettit suggested that it either occurs naturally in *A. convoluta*, or is selectively sequestered or transferred from *B. neritina*, or else, there is a common food source. The case of *Lissodendoryx* is interesting, because it was collected in Bahia de Kino, Sonora, Mexico, on the same expedition as the highly active *Bugula neritina* collections. The sponge *Lissodendoryx isodictyalis* had previously been collected several times for the NCI in Floridian and Caribbean waters and had proven totally inactive and nontoxic. After questions of correct identity were resolved, it was suspected that the project might be something of a red herring and work by Pettit et al. confirmed that there was a substantial amount of *Bugula neritina* (2–5%) in the cavities of this sponge and that the active components of *L. isodictyalis* were in fact bryostatins [52]. Two new bryostatins, A and B, were found in trace amounts in addition to the already known bryostatins 4, 5, 6, and 8. The tunicate *Aplidium californicum*, likewise collected from Bahia de Kino, Sonora, also contained bryostatins 4 and 5. Close examination of the voucher specimens demonstrated the presence of significant amounts of *Bugula neritina* [53].

It is noteworthy that the initial isolations of bryostatins 1 and 2, and the first in vivo tests on pure material were carried out in the Spring of 1981. Bryostatin 1 was evaluated in three experiments against P388 leukemia on a daily injection protocol for 9 days with the results shown in Table 4. These data show that: (1) bryostatin 1 had activity (ILS $>25\%$) over a broad dose range from 155 down to 3 μg/kg/day; (2) it was a highly potent material; and (3) there was substantial but unspectacular activity as no animals were cured of tumor. Similar data was noted for bryostatin 2 (Table 4), which was however less potent and resulted in a somewhat smaller increase in life span. Actually, the antitumor data at that stage looked quite similar

Table 4. Activity of bryostatins 1 and 2 against P388 leukemia on a daily schedule for 9 days

Compound	Expt. #	Dose μg/kg/day	% ILS
Bryostatin 1	08/3810	620	toxic
		310	toxic
		155	79
		77	63
Bryostatin 1	08/3866	70	96
		35	86
		18	52
Bryostatin 1	08/3914	15	60
		7	41
		3	36
Bryostatin 2	08/3798	1200	toxic
		600	toxic
		300	60
Bryostatin 2	06/J10387	600	73
		300	56
		150	63
		75	49
		37.5	44

to data for certain phorbol-like compounds, such as gnidimacrin and related daphnane esters that had previously been isolated by Kupchan and coworkers [54, 55].

6.3.2 Detailed Biological Evaluation

At that point it had to be ascertained, whether the bryostatins had activity in tumor models other than P388 leukemia, and if so, at what level. The evaluation method in use at that time at the NCI consisted of testing those compounds with activity in the P388 in vivo prescreen in a panel of more resistant tumors. The results for bryostatins 1 and 2 are presented in Table 5. This testing was largely done in 1981 and 1982 and the opinion of NCI staff members at that time was that these data were unimpressive, as the greatest activity was found only in the most sensitive model (P388 leukemia) and that there was no basis for placing the compounds into preclinical development. Further testing remained to be done, however; in 1981 the NCI contracted a recollection of 1,000 gallons of *B. neritina* to the Pacific Biomarine Company. This was worked up through the early stages of extraction, solvent partitions and first stage chromatography in 1983 at Polysciences, Inc., under contract to the NCI. A flow sheet for the initial stages is shown in Fig. 8. Chromatography of 1.8 kg of active partition fraction on silica gel using increasing amounts of methanol in dichloromethane yielded bryostatin-rich fractions, which

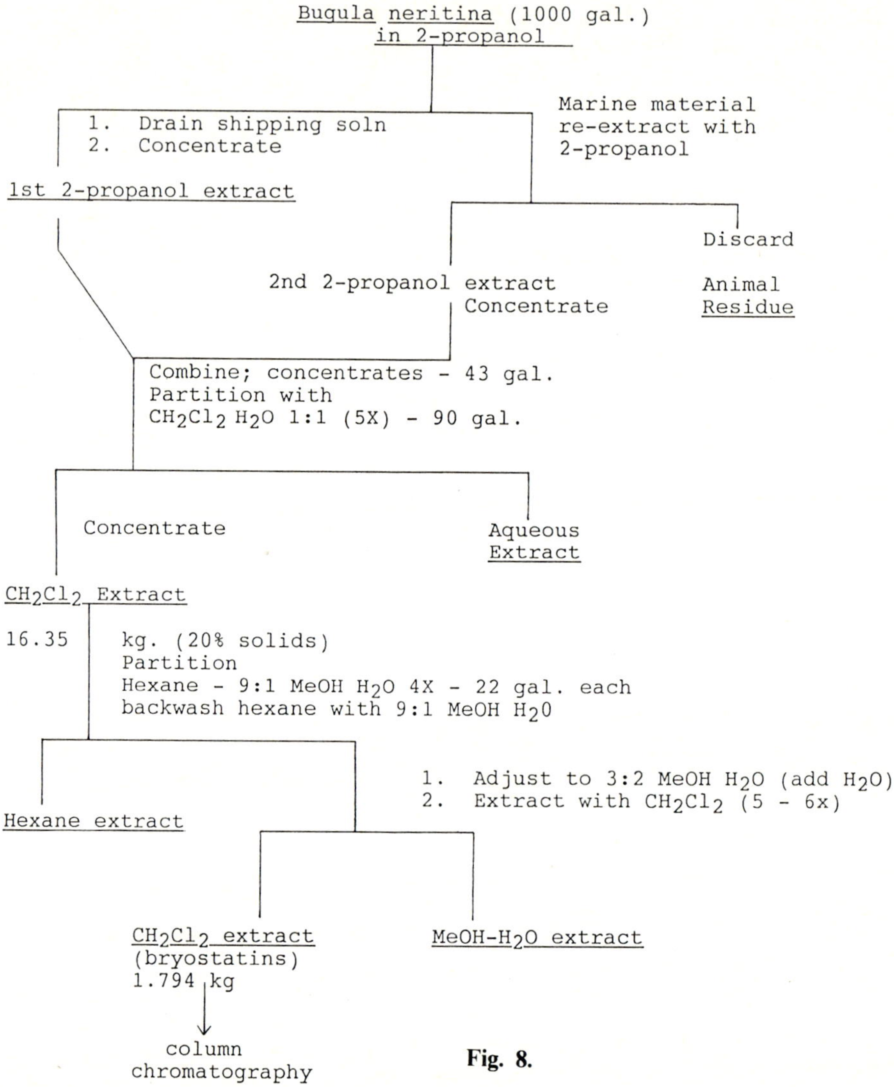

Fig. 8.

were transferred to Professor Pettit's group, who isolated 630 mg of bryostatin 1 and 588 mg of bryostatin 2. It is uncertain, however, whether these yields are representative, since the organism was collected in 1981 and was stored in barrels in isopropanol until 1983, when it was processed.

Additional antitumor testing at the NCI was not especially encouraging. Since the pattern of antitumor activity exhibited by the bryostatins, clearly resembled that of phorbol-type compounds, there was also concern over possible promotion of carcinogenesis. Up to that time (1981), no studies of the mechanism of action of

Table 5. Summary of NCI tumor panel testing of bryostatins 1 and 2

Model	Activity*	
	Bryo. 1	Bryo. 2
B16 melanoma	neg	neg
CDF_1 mammary	neg	neg
CX-1 colon xenograft	neg	N.T.
colon 38 carcinoma	neg	neg
L1210 leukemia	+	neg
LX-1 lung xenograft	neg	N.T.
Lewis lung carcinoma	neg	neg
MX-1 mammary xenograft	neg	neg
M5076 sarcoma	+	+
P388 leukemia	++	++

* Activity ratings are: neg = inactive; + = weak to modest reproducible activity; ++ = strong activity; +++ = curative activity, N.T. = not tested

bryostatin had been performed. In the initial paper, which disclosed the structure, Pettit implied that the bryostatins might be ionophores. Structural resemblance to compounds like aplasmomycin [31] is apparent, which supports that hypothesis. The first mechanistic study was reported in 1985 by Berkow and Kraft [56], who found that bryostatin 1 binds to the phorbol receptor (equivalent to calcium, phospholipid-dependent protein kinase, protein kinase C). These workers also reported that bryostatin blocks the phorbol-induced differentiation of HL-60 promyelocytic leukemia cells [57], an important observation that established that the bryostatins are not simple phorbol analogs. This work has been extended in Blumberg's laboratory, where it has been demonstrated that the effects of bryostatin cannot be accounted for by simple competition with phorbol esters at the phorbol binding site, and that there must be a second site of action. Studies by Blumberg, Pettit and, coworkers with radiolabelled bryostatin 4 suggest that bryostatin binding may anchor the protein kinase C at the first location of recognition, whereas phorbol esters may permit redistribution of protein kinase C with time [58]. Hennings et al. have studied tumor promotion by bryostatin 1 and found that not only was it not a promoter in contrast to the phorbol esters, but that it in fact suppressed promotion by phorbols [59]. The importance of protein kinase C in signal transduction in cells, and evidence that this enzyme is stimulated by certain oncogenes has made this a prime research target in the last few years. Protein kinase C and the associated diacylglycerol–phosphoinositol pathway is clearly a major signalling and regulatory system at the cellular level.

On the basis of the early work with the phorbol receptor, Pettit encouraged the Cancer Research Campaign in Great Britain to consider development of bryostatin 1, and in Augsut, 1985, this organization agreed to begin development. However, only minimal work was possible until 1987–88, when more material was made available from another collection that was processed by Pettit. Meanwhile, mechanistic studies were proceeding at NCI and elsewhere and the question was discussed whether NCI should now reconsider bryostatin 1. While the protein kinase C data were interesting, they did not alter the modest level of in vivo antitumor activity. The decision might have been negative, except for two new pieces of data. First, the NCI had just developed a new screening concept that was based on searching for disease-specific cytotoxicity with panels of various human cell lines. The objective was to find drugs with highly selective specificity so that toxicity to nontarget tissues could be greatly reduced. Bryostatins 1 and 2 showed very strong specificity for the leukemia cell lines with differences in IC_{50} values for growth inhibition between sensitive and resistant lines of 1,000 fold or more (see Fig. 9). This was the greatest differential toxicity of any drug screened to date. Second, work by May, Sharkis, and coworkers at Johns Hopkins University showed that bryostatins are stimulators of human hematopoietic progenitor cells [60]. The bryostatins directly stimulate bone marrow progenitor cells to form colonies and they functionally activate neutrophils. Thus they mimic effects of colony stimulating factors (CSFs), such as the human granulocyte-macrophage CSF. This was an amazing result, since a great majority of cytotoxic agents used in cancer chemotherapy are rather toxic to bone marrow elements, and bone marrow suppression is a major frequently encountered toxicity in cancer chemotherapy. Most antileukemic agents are toxic to many rapidly dividing cells, but the bryostatins clearly are exceptional in that they are selective for leukemias. This early example illustrates the power and utility of the new NCI screening system in selecting promising candidates for clinical trials.

Potential roles for bryostatin in cancer therapy include: (1) use in combination chemotherapy, where its bone marrow CSF-like effects could counteract the myelosuppression of existing agents; (2) in maintenance therapy between courses of conventional chemotherapy, where its modest antitumor activity and its capability to stimulate hematopoiesis would be useful.

Based on this new information the NCI decided in January, 1988, to place bryostatin into preclinical development. The first concern was supply of material, as only micrograms were available at the NCI and Pettit's limited supply was committed for the studies by the Cancer Research Campaign in Great Britain. A contract was negotiated with Marinus, Inc., Long Beach, California, to collect 11,000 gallons of organism (200 55-gallon drums) packed in isopropanol. This work began in February, 1988, and was completed in July, 1988. This material is being extracted and the extract will be processed at the Frederick Cancer Research Facility. Based on current data, the expected yield is a minimum of 10 g of bryostatin 1.

6.3.3 Bulk Production

While we recognize that some novel bioactive metabolites from marine organisms will be amenable to practical laboratory synthesis, many will not. The following discussion points up some considerations and problems in scale-up with the

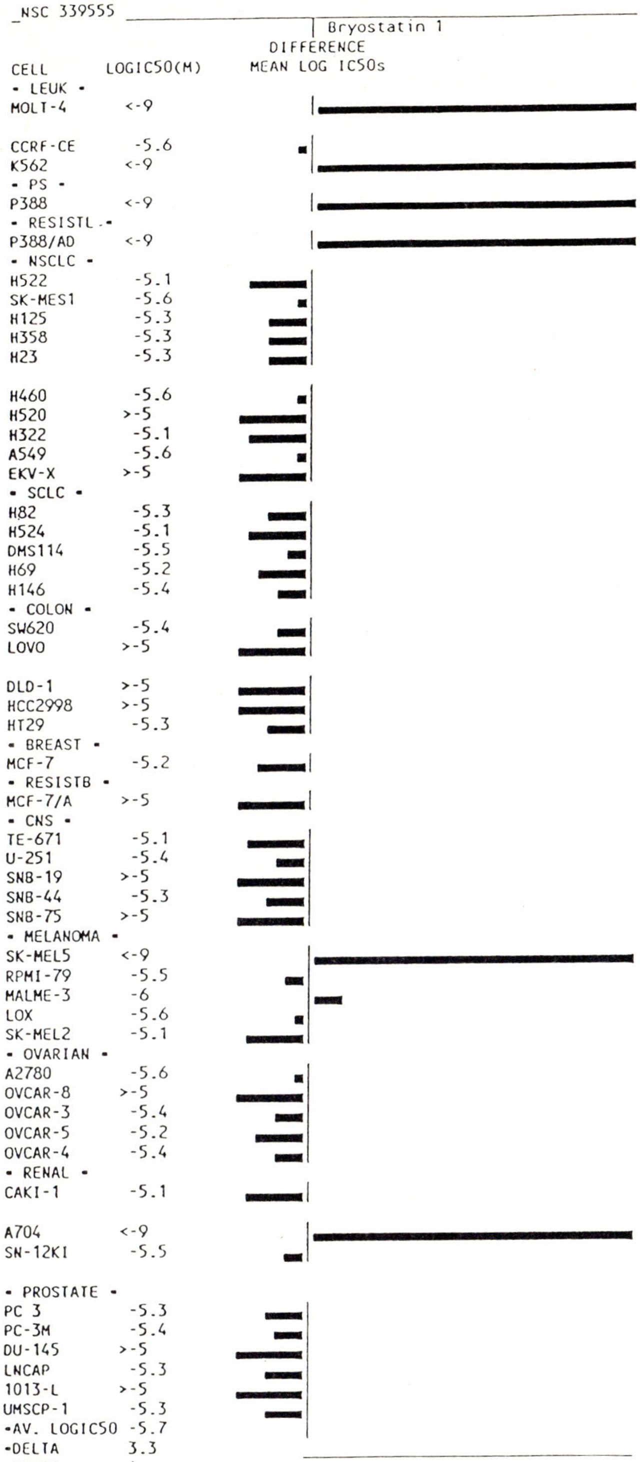

Fig. 9. In vitro cell cytotoxicity of bryostatin 1

isolation of bryostatins from *Bugula neritina* as a prominent example. A major task lies in redesigning the previously used laboratory process for scaled-up production under GMP (good manufacturing practices) conditions. The bench chemist does not have to be concerned with such factors as cost of adsorbent, the exact adsorbent:charge ratio, the cost of waste disposal, efficient recycling of solvents, the maximum through-put of solvent in a large column per day, and solvent requirements for a given flow rate. These and other criteria are critical for work on pilot plant scale. Not only are recovery yield and purity of product important, but also the efficiency and costs of the process, as is safety of the workers. There is always a need to restudy laboratory processes extensively for scale-up feasibility, efficiency, cost and safety; in most cases the laboratory process can be considerably shortened. This in no way makes light of the capabilities of the bench chemists, but rather is a reflection of the reality that in the original bioassay-directed isolation the chemist has little or no idea of the chemical nature of the active substance and must treat it very gently and cautiously which accounts for the separation methods that are used. Once the chemistry of the compound is known, the development chemist in redesigning the process must first consider chemical "handles", which can be used to expedite separation. For example, if the active compound is a carboxylic or sulfonic acid or a basic amine, it may be easy to purify the active material 10 to 50 fold by use of acid-base shakeouts or ion exchange chromatography. Any chemical reaction conditions which do not affect the desired compound can be used to convert undesired constituents to more easily removed derivatives. Such chemical transformations, for reasons of cost and safety, would not generally be employed at early stages but may be extremely useful in final stages of purification in order to convert a closely related impurity to a more or less polar derivative that is readily separable from the desired compound. Typical reactions might include esterification, hydrolysis, oxidation, or reduction of a functional group in the undesired compound, which is either absent or unreactive in the desired compound. This type of separation enhancement may be necessary in scale-up, because difficult separations achievable on analytical HPLC, for example, are generally not achievable on the large columns needed in scale-up. The reverse approach of chemical transformation, derivatization of the desired compound to facilitate separation, has also been used, although it requires extra steps to reconvert to the original compound and to purify the reaction mixture. A good example of this process was the anitumor amino acid acivicin, which was extremely water-soluble and difficult to separate from very polar impurities. Acivicin was converted to a *t*-BOC derivative, which was then extracted into an organic solvent, purified, and quantitatively hydrolyzed to the original amino acid [61].

Another key component in process development for scale-up production is a fast and specific assay for the desired compound. This requirement can often be met by an HPLC assay, but in many cases this is unsatisfactory in the early stages of the isolation process because the desired compound cannot yet be seen on an HPLC trace. At times, it cannot be quantitatively assessed due to its low concentration in the extract, its similarity to related components, its lack of strong UV, visible, or fluorescent absorption for detection, or high background absorption from other components in the extract. Generally, one can use a bioassay-directed isolation to monitor process development, but caution must be exercised, when that assay is a fairly general one, such as cytotoxicity or antimicrobial activity. Many extracts will contain a number of active compounds of chemically and mechanistically diverse types, which produce the same net effect of cell kill or growth inhibition. One can circumvent this problem by using multiple bioassays with different specificities, which will discriminate among the desired compounds. In the case of *Bugula neritina*, we might have tested all the early process development fractions in both the P388 and KB cell lines. Fractions showing similar cytotoxicity towards both cell lines could be presumed to be non-bryostatin-like, since the P388 line is at least 1,000-fold more sensitive to bryostatins than is the KB cell line. However, research on the mechanism of bryostatin action in Blumberg's laboratory, which showed high specificity of bryostatins for protein kinase C led to our collaboration, which resulted in a phorbol displacement assay. This was used to rapidly assess total bryostatin-like activity in the large number of fractions that were studied during process development by Beutler and coworkers at the Frederick Cancer Research Facility [62]., This assay is ideal for the NCI's purpose, since it detects all bryostatins which are bioactive in this system, as it is of interest to isolate as many bryostatins as is reasonable for further mechanistic and structure-activity research. The structural similarities of most of the bryostatins, which are esters of the same alcohol, also may offer the possibility of chemical conversion to produce additional amounts of bryostatin 1. Had the situation been different,where only one of a group of closely related compounds had been desired, a more specific assay, such as a radioimmunoassay, might have been developed. A future possibility for large-scale production of bryostatin 1 may be by culturing *Bugula* using aquaculture methods. In that case, the desired result will be achieved by selecting strains or conditions that produce predominantly bryostatin 1, rather than total bryostatins. While a bryostatin binding assay may be useful as a preliminary screen to determine which cultures produce bryostatins, a more definitive assay will be required to select the best cultures for further optimization and production.

Binding assays can be extremely useful, but caution must be exercised. Various active analogs have different binding affinities, so that the total activity numbers do not relate directly to mass of active components, but rather to sum of the products of mass times binding constant of the active components. Thus, presence of a minor component with a high binding constant will result in overestimation of the amount of active major component. This type of assay is most convenient, when the major sought-after compound has a higher binding constant than its minor congeners, thus making only a small net contribution to activity of the minor constituents. Before beginning to use such an assay for optimization, one should determine the binding constants of all co-occurring related compounds.

In addition to considering modifications of the original isolation scheme for process development, as discussed above, one should also study possible use of affinity chromatography, whenever a biochemical mechanism of action of the active compound is known. Affinity chromatography is a powerful method for concentrating all materials in an extract which possess similar biological receptors or targets, thereby readily removing nearly all other impurities. For example, several years ago the process development group at the Frederick Cancer Research

Facility developed a method for isolating the enzyme asparaginase from complex fermentation broths of *Erwinia caratovora* [63]. Asparagine which is the normal substrate for the enzyme, was covalently bonded to a solid substrate of Sepharose CL-4B. When a crude fraction from the fermentation broth was passed through the immobilized asparagine column at a rapid flow rate, only the desired enzyme asparaginase was held tightly. After thorough washing of the column, a change in buffer resulted in elution of highly concentrated asparaginase of good purity, ready for final purification steps. In the above format, affinity chromatography is used to isolate an enzyme or protein; this is a well-established technique for purification of biological receptors. A reversal of the assay can be equally useful in some instances, where the receptor or binding component is immobilized. However, the number of receptor sites or binding sites in a biological preparation may be relatively small. Hence it is not feasible to acquire enough material for an affinity column. For isolation of the bryostatins, affinity chromatography using immobilized protein kinase C was considered, but the enzyme has poor stability characteristics and can only be isolated in very small amounts from rat membrane preparations. The number of rats, the amount of labor and time involved made such an approach unrealistic in this case, but one could certainly envisage that, if protein kinase C were cloned and expressed in a suitable microorganism to make it readily available, affinity chromatography could be used for bryostatin isolation from *Bugula neritina* in the future.

7 Conclusions

We have attempted to give the non-expert reader an overview of discovery and development of antineoplastic agents from natural sources, with particular emphasis on the marine environment. While there are many important points in this area, the two most critical components are biological assays and knowledge of the ecology of the producing organism.

Choice of bioassays for prescreening, screening, and in monitoring isolation will dictate what kinds of substances can be discovered and what their biological properties will be. Once substances of interest are identified, it is critically important to be able to develop them for human trials through bulk production. Bulk production requires that we understand which organism or which combination of different organisms produces the compounds, at what stages they are produced, and which ecological stressors or dietary factors affect production.

8 References

1. Boyd MR, Shoemaker RH, McLemore TL, Johnston MR, Alley MC, Scudiero DA, Monks A, Fine DL, Mayo JG, Chabner BA (1988) in: Roth JA, Ruckdescher JC, Weisenburger TH (eds) Thoracic Oncology. Saunders, New York, Chap 51
2. Waksman SA, Woodruff HB (1940) Proc. Soc. Exptl. Biol. Med. 45:609
3. Meienhofer J, Atherton E (1973) Adv. Appl. Microbiol. 16:203

4. Bunge RH, McCready DE, Balta LE, Graham BD, French JC, Dion HW (1979) Devel. Indus. Microbiol. 20:393
5. Goldin A, Serpick AA, Mantel N (1966) Cancer Chemother. Rep. 50:173
6. Goldin A, Schepartz SA, Venditti JM, DeVita VT Jr (1979) in: DeVita VT Jr, Busch H (eds) Methods in Cancer Research. XVI A Academic Press, New York, p 165
7. Johnson RK, Goldin A (1975) Cancer Treat. Rev. 2:1
8. Venditti JM, Wesley RA, Plowman J (1984) Adv. Pharmacol. Chemother. 20:1
9. Boyd MR (1986) in: Frei EJ, Freireich EJ (eds) Accomplishments in Oncology. Lippincott, Philadelphia p 68
10. Boyd MR (1985) Fourth World Conference Lung Cancer: Mini Symposium Proceedings, Toronto, 1985
11. Mosmann T (1983) J. Immunol. Meth. 65:55
12. Alley MC, Scudiero DA, Monks A, Hursey ML, Czerwinski MJ, Fine Dl, Abbott BJ, Mayo JG, Shoemaker RH, Boyd MR (1988) Cancer. Res. 48:589
13. Faulkner DJ (1984) Nat. Prod. Rep. 1:251
14. Faulkner DJ (1984) Nat. Prod. Rep. 1:551
15. Clement JJ, Burres N, Blunt JW, Munro MHG, Higa T (1988) Proc. Amer. Asso. Cancer Res. 29:1285A
16. Perry NB, Blunt JW, Munro MHG, Pannell LH (1988) J. Am. Chem. Soc. 110:4850
17. Rinehart KL Jr, Gloer JB, Cook JC, Mizsak SA, Scahill TA (1981) J. Am. Chem. Soc. 103:1857
18. Chun HG, Davies B, Hoth D, Suffness M, Plowman J, Flora K, Grieshaber C, Leyland-Jones B (1986) Invest. New Drugs 4:279
19. Rinehart KL, Kishore V, Nagarajan SW, Lake RJ, Gloer JB, Bozich FA, Li KM, Maleezka RE Jr, Todsen WL, Munro MHG, Sullins DW, Sakai R (1987) J. Am. Chem. Soc. 109:6846
20. Pettit GR, Kamano Y, Brown P, Gust D, Inoue M, Herald CL (1982) J. Am. Chem. Soc. 104:905
21. Pettit GR, Kamano Y, Herald CL, Tuinman AA, Boettner FE, Kizu H, Schmidt JM, Baczynskyj L, Tomer KB, Bontems RJ (1987) J. Am. Chem. Soc. 109:6883
22. Pettit GR, Herald CL, Doubek DL, Herald DL, Arnold E, Clardy J (1982) J. Am. Chem. Soc. 104:6846
23. Smith JB, Smith L, Pettit GR (1985) Biochem. Biophys. Res. Comm. 132:939
24. Pettit GR, Herald CL, Kamano Y, Gust D, Aoyagi R (1983) J. Nat. Prod. 46:528
25. Pettit GR, Herald CL, Kamano Y (1983) J. Org. Chem. 48:5354
26. Pettit GR, Kamano Y, Herald Cl, Tozawa M (1985) Can. J. Chem. 63:1204
27. Gerwick WH, Fenical W, van Engen D, Clardy J (1980) J. Am. Chem. Soc. 102:7991
28. Moore RE (1982) Pure Appl. Chem. 54:1919
29. Mynderse JS, Moore RE, Kashwagi M, Norton TR (1977) Science 196:538
30. Mynderse JS, Moore RE (1978) J. Org. Chem. 43:2301
31. Okami Y (1982) Pure Appl. Chem. 54:1951
32. Proctor NH, Chan SL, Trevor AJ (1975) Toxicon 13:1
33. Bates HA, Kostriken R, Rapoport H (1978) Toxicon 16:595
34. Murakami Y, Oshima Y, Yasumoto T (1982) Bull. Jpn. Soc. Sci. Fish. 48:69
35. Yasumoto T, Sieno N, Murakami Y, Murata M (1987) Biol. Bull. (Woods Hole) 172:128
36. Yotsu M, Yamazaki T, Meguro Y, Endo A, Murata M, Naoki H, Yasumoto T (1987) Toxicon 25:225
37. Tachibana K, Scheuer PJ, Tsukitani Y, Kikuchi H, van Engen D, Clardy J, Gopichand Y, Schmitz FJ (1981) J. Am. Chem. Soc. 103:2469

38. Schmitz FJ, Prasad RS, Gopichand Y, Hossain MB, van der Helm D, Schmidt P (1981) J. Am. Chem. Soc. 103:2467
39. Keough MJ, Chernoff H (1987) Ecology 68:199
40. Kitamura H, Hirayama K (1984) Bull. Jpn. Soc. Sci. Fish. 50: 973
41. Keough MJ (1984) Evolution 38:142
42. Kitamura H, Hirayama K (1984) Bull Jpn Soc Sci Fish 50:1
43. Vail LL, Wass RE (1981) Aust J Mar Freshwater Res 32:639
44. Keough MJ (1986) Ecology 67:846
45. Kitamura H, Hirayama K (1985) Bull. Jpn. Soc. Sci. Fish. 51:1401
46. Pettit GR, Kamano Y, Herald CL, Tozawa M (1984) J. Am. Chem. Soc. 106:6768
47. Pettit GR, Kamano Y, Herald CL (1986) J. Nat. Prod. 49:661
48. Pettit GR, Kamano Y, Herald CL (1987) J. Org. Chem. 52:2848
49. Pettit GR, Leet JE, Herald CL, Kamano Y, Boettner FE, Baczynskyj L, Nieman RA (1987) J. Org. Chem. 52:2854
50. Pettit GR (1988) personal communication
51. Pettit GR, Kamano Y, Aoyagi R, Herald CL, Doubek DL, Schmidt JM, Rudloe JJ (1985) Tetrahedron 41:985
52. Pettit GR, Kamano Y, Herald CL, Schmidt JM, Zubrod CG (1986) Pure Appl. Chem. 58:415
53. Pettit GR, Leet JE, Herald CL, Kamano Y, Doubek DL (1986) J. Nat. Prod. 49:231
54. Kupchan SM, Shizuri Y, Murae T, Sweeny JG, Haynes HR, Shen MS, Barrick JC, Bryan RF (1976) J. Am. Chem. Soc. 98:5719
55. Kupchan SM, Sweeny JG, Baxter RL, Murae T, Zimmerly VA, Sickles BR (1975) J. Am. Chem. Soc. 97:672
56. Berkow RL, Kraft AS (1985) Biochem. Biophys. Res. Commun. 131:1109
57. Kraft AS, Smith JB, Berkow RL (1986) Proc. Natl. Acad. Sci. USA 83:1334
58. Blumberg PM (1988) Cancer Res. 48:1 and reference therein
59. Hennings H, Blumberg PM, Pettit GR, Herald CL, Shores R and Yuspa SH (1987) Carcinogenesis (Lond.) 8:1342
60. May WS, Sharkis SJ, Esa AH, Gebbia V, Kraft AS, Pettit GR, Sensenbrenner LL (1987) Proc. Natl. Acad. Sci. USA 84:8483
61. Martin DG, Biles C, Mizsak SA (1981) J. Antibiot. 34:459
62. Beutler J (1988) personal communication
63. Lee S-W, Ross JT, Gustafson ME, Wroble MH, Muschik GM (1986) Appl. Biochem. Biotech. 12:229

Author Index Volumes 1–3

Subject Index